ACE THE PMP® EXAM: 50 CRITICAL PATH METHOD (CPM) EXERCISES TO HELP YOU PASS YOUR PMP® EXAM

By
Glen Ford, PMP

Published By

Mississauga, Canada

Published by TrainingNOW, Mississauga & Oakville, Ontario, Canada
http://www.TrainingNOW.ca
http://www.LearningCreators.com

ISBN (Kindle Edition): xxx-x-xxxxxxxx-x-x
ISBN (Print Edition): 978-1-5431111-9-4

PMP®, *PMBOK® Guide*, PMI® and several other related registered marks used throughout this book are registered marks of the Project Management Institute, Newton Square, PA, USA.

R20170309.205600

For my long-suffering wife, Lisa

who puts up with me

despite everything.

I really do appreciate you.

Acknowledgements

No book is ever created just by its author. Many people had a hand in this book's creation. I would especially like to thank the following people.

To my wife Lisa. Who has put up with me through the long days (and sometimes nights) as I learned to perfect my art. And then put up with me while I turned that art to building companies and finally to writing and teaching. Who sometimes fed me, and often carried me. I may not say it, but I really do know what you've gone through for me.

To my son Dafydd and daughter Solenne, who have taught me the true meaning of herding cats. And the joy to be found in helping people to achieve their dreams. I am very, very proud of you.

To my late mother and father who taught me, guided me and pushed me. I owe you more than I can ever say.

To my students who have graciously listened to me drone on, laughed at my poor jokes, and forgiven me when my jokes have gone over the line. You've taught me as much as I've taught you.

To my students and others who read my early drafts. Your corrections and suggestions (and yes, reviews) have helped immensely.

I hope that you enjoy this book and that it proves to be useful to you. Project Management is a profession that is frustrating in its complexity and limited in its appreciation. Your efforts in achieving certification helps the entire profession move forward in becoming recognized for the value it brings. I hope this book helps you to achieve your project management goals in some small way.

Could you take five minutes to help in return?

Other readers rely on your opinions when choosing which books to spend their precious time reading. Writers like me rely on your feedback to improve their writing and to know what topics to write about next.

If you enjoyed this book, please express your rating and opinions on Amazon. Amazon provides a facility for rating and reviewing on the page where you purchased this book. You can reach the review page by going to the bottom of http://www.amazon.com/dp/154311119X and selecting the 'write a customer review' button.

On the other hand, if you didn't enjoy this book or if you want me to write about other topics, please let me know directly. You can do so through the contact form available at my publisher's site http://TrainingNOW.ca or the one at my own website http://GlenDFord.com/. While you are at my site, feel free to check out my blog. It contains many of my thoughts on project management, innovation and business management for entrepreneurs. You can also join my email list. If you do, I'll keep you appraised of new books and other news for project managers.

Thank You and Enjoy!

Glen Ford

Table of Contents

Chapter 1: Introduction

So you've decided to stand for your Project Management Professional or PMP® certification, have you? Good for you. You've made a decision that will enhance your career for many years to come. No one will ever claim that the exam you are about to write is an easy one. Nor that the qualifications you require to stand for the exam are light. However, the PMP® certification from PMI® is the premier qualification for Project Managers. It is well worth the effort you are about to spend to obtain it.

Unlike most of the other project management certifications, the PMP® is attempting to measure your understanding of the underlying theory of project management. While it is not the last certification you can expect to be required to have, it is the most important. If approached with the right frame of mind, the PMP® and the PMBOK® GUIDE will explain to you why the methodologies behind other certifications work. And why certain methodologies fail when applied in the wrong situation.

The PMP® is a required certification in many markets, especially in North America. Without it, you will be unlikely to find any position as a project manager. Employers are demanding it. Even when they don't really understand what it signifies. Even though they can expect to pay more for people who hold it. All they know is that without those three letters, the project managers they hire will not be as good.

While the validity of those opinions may be arguable, there are two facts that are not. PMI's studies have indicated that project managers with a PMP® tend to earn 20% more than their equivalents without a PMP®. And the difference is increasing (16% in 2011, 17% in 2015, and 20% in 2016). Second, when more than one third of an organization's project managers are PMP® certified, the success rate for projects increases significantly (roughly 20%). So both the individual and the organization gain when project managers are PMP® certificated.

Part of the requirement for taking your PMP® is that you have extensive experience managing and directing projects. Depending on your education, you need a minimum of either three or five years working full-time as a project manager. This means you probably have at least ten years' experience overall. This experience requirement is intended to ensure that you really understand the knowledge domains involved in project management. After all, the PMBOK® GUIDE is only a guide to the knowledge required. At one time, this would have been reasonably true with a few notable exceptions.

However, as the discipline has expanded and become more important to more organizations, the technology supporting project management has also improved. The size and type of organizations supporting project management have also changed in that period. The result is that we've developed holes in our experience. Things that we need to know are hidden from us. Tools spit out answers without the background information to judge accuracy. Processes that were performed properly are now performed haphazardly by organizations. Candidates no longer work in organizations that have a project management methodology. Experience is no longer a guarantee of breadth or depth of knowledge. Experiences that old fogeys like me took for granted, are no longer provided to the younger generation of project managers.

This isn't necessarily a bad thing. After all, the use of technology means that we can no longer justify hiding behind piles of reports, playing with figures. Instead, we are required to be out with our people. We're out doing what really matters. We're out actually performing project management rather than performing project reporting. Unfortunately, we are developing holes in our knowledge.

Over the last few years, as a trainer of project management skills, I have experienced this directly. Many of my students have never worked with a professional purchasing group. Many have never scheduled a project manually. Many have never formally managed risk events. I've yet to meet a student who has used Earned Value Management (EVM) in their practical life. Very few have used a professional accountant. Many have never led a project outside of their technical background.

This book series is all about filling those holes.

Each book in this series will focus on a single topic my students have indicated they will need further practice in. I'll provide a series of cases and questions that will help the potential PMP® candidate practice skills that they may not need to use in their jobs. Skills that may be provided by machines toiling in secret. Skills that are nonetheless critical to being considered a professional.

The books all begin with a short description of the process involved. In most cases, this will be a review rather than actually teaching the topic. The PMBOK® GUIDE and your formal project management courses will have taught the material. I'm not going to reteach material you already know. Although I do encourage you to read this section. After all, I may have the insight that will "turn on the light bulb" for you.

The second section is the important section. It will consist of questions and exercises to allow you to practice the skills. I recommend that you go through each question, write the answer on a separate sheet of paper, and then look at the answer. For most of the books, you'll find the answer immediately following the question. If you got the right answer, fine. Go on to the next question.

If you did not get the right answer then look at the explanation. Reread the explanation of how to do the task. Determine where you went wrong. In many cases, the answer is self-explanatory. However, it may not be enough. That's why I have explained how I got the answer I did. If a reference to PMI's Guide to the PMBOK® is appropriate, you'll find it there.

The questions are, generally speaking, in order of increasing difficulty. I encourage you to go through all the questions. However, the reality is that the exam itself has a limit to how hard the question can be. After all, it must be answered in one or two minutes. My book should provide questions that are more difficult that you'll find on the exam. I've done this on the philosophy that underprepared means you may fail. Over-preparation means that correctly answering simpler questions is easy.

Good luck and have fun.

Chapter 2: The Critical Path Method

The Critical Path Method or CPM was developed in the late 1950's by Morgan R. Walker of DuPont and James E. Kelley, Jr. of Remington Rand (which became Unisys eventually). CPM was based on the techniques used by DuPont during 1940's. More specifically the techniques were those used on the Manhattan Project. So for those of us raised in the Post-Atomic age, we can blame CPM for the bomb, hippies, and the peace movement.

Although CPM was developed prior to (or maybe congruent with) Dr. John Fondahl's Precedence Diagramming Method, modern versions of CPM are based on the PDM chart. The original activity-on-arrow (AOA) diagram has largely gone the way of the Ford Model T. It has been replaced with the activity-on-node (AON) diagram of PDM which is much more computer friendly.

The PDM chart is used to show how tasks (shown as blocks or circles known as nodes) are interconnected. Effectively it shows the logic network or sequence of your project tasks. By following the network diagram from beginning to end you can determine the sequence your project tasks need to be done in. You can also determine which tasks can be performed in parallel, and which ones need to be sequential. Each of these sets of tasks is a path through your PDM chart.

Critical Path Method uses this sequence to determine which path is the most critical. It is that path (or paths) that determines how long the project will take. It also is the most likely path to fail and cause the project to be late. So this path becomes the major focus for a project manager. This is why we refer to the path as the Critical Path. CPM then uses that path to determine when tasks must be performed in order to complete the project within the desired time period.

There are a number of variations on the process of developing a critical path. All are correct and acceptable. All are basically the same thing. Some condense steps. Some expand steps. Some start with the first task and some with the start task. Ultimately, all work the same way and most are "correct". It's mostly a matter of how the presenter chooses to connect the dots. In my case, I'm going to break the process down to its simplest parts.

Before we start, we need to make a decision. Specifically, we need to decide the period of our diagram. Is it one week? Is it one day? Is it one month? Is it one year? Is it one shift? While we could (theoretically) map multiple periods, for simplicity's sake, we really only want to use one period. Whatever period we want is fine. But just one per diagram. Otherwise, we'll confuse ourselves. Leave that for Microsoft Project or when we are experts.

In the rest of this book, I am always going to set the period to one day. So if I say one day, feel free to substitute the word period. Or whatever period you feel more comfortable using. It's just that it's easier to understand if I use a real period of time, rather than saying period all the time. And it's also shorter to type.

Now that we've made that decision, we can start learning about the process. In all cases we start with a simple box (see figure 1). This box represents the task we are going to perform. It has a description or title or name. It has an id. It has a duration. It has a start and it has a finish. When it starts and when it finishes is what we are going to determine using CPM. And in fact, it has a best-case and a worst-case scenario. The best case we call the early start and the early finish. The worst case we call the late start and late finish. The difference between the two is how much time we can play with (i.e. delay) the task without causing the project to be late. We call this value the float. All of those calculations are the focus of CPM.

Figure 1

The box itself represents the activity. On the right hand, we have the start. On the left, we have the finish. We will always start at the beginning of the day. And we will always finish at the end of the day. That means that the next task will always start one day later than we finish this task. So for example, if I finish task "First Task" at 11:59.59 pm on the third day, then the next task, "Second Task", will start at 12:00.01 am the fourth day. Makes sense? Although this is a very simple concept, it can be the source of much gnashing of teeth by students. And dollars to donuts, you will make the mistake at least once in the exercises. Just don't do it on the exam.

The other item to notice with the task box (or node) is that the early dates are on the top, and the late dates are on the bottom. We call filling in the top (or earliest we can do the task), the forward pass. Calculating the bottom figures (i.e. the latest we can do the task without the project being late), we call the backward pass.

The basic process consists of:

1. Determine the paths
2. Determine the longest paths - the Critical Paths
3. Calculate the early dates for the Critical Path items
4. Record the late dates and the floats for the Critical Path items

5. Calculate the early dates for each of the non-critical path items
6. Calculate the late dates for each of the non-critical path items
7. Calculate the float(s) for each of the non-critical path items.

We'll go through each of those now.

Step 1 - Determine the paths

The first thing we need to do is simply to identify the paths through the network diagram. You do that by following the arrows through the CPM chart. Remember that at this stage, there is no difference between a PDM chart and the CPM chart. Technically, PDM allows nodes to be circles and CPM requires boxes as its nodes. But effectively, they are the same thing at this point.

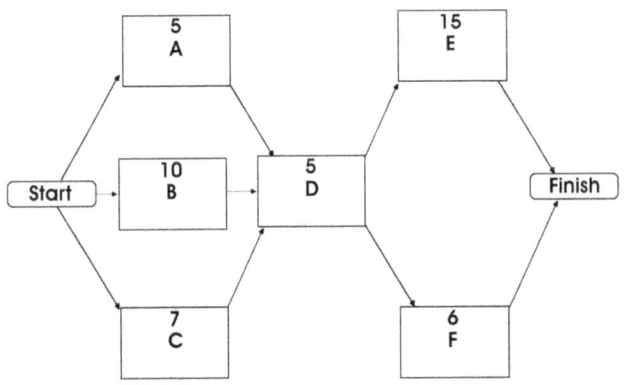

Figure 2

You can see an example of this chart in Figure 2.

If I trace the paths through Figure 2, I get the following six paths:

- start-A-D-E-finish
- start-A-D-F-finish
- start-B-D-E-finish
- start-B-D-F-finish
- start-C-D-E-finish
- start-C-D-F-finish

Note that in most cases, when identifying the path, you will drop the start and finish. So start-A-D-E-finish is usually just named as A-D-E. I broke the normal rules in order to make the path clearer to you.

Step 2 - Determine the longest paths - the Critical Paths

Now that we have the paths, we need to determine which of the paths is the Critical Path. To do this, simply add the durations for each of the tasks in the path. You can find the duration in the middle position of the top row. Normally, the duration is directly above the Id. or title. The path or paths with the longest duration is (or are) the Critical Path.

So using Figure 2 we get the following:

- A-D-E = 5 + 5 + 15 = 25
- A-D-F = 5 + 5 + 6 = 16
- B-D-E = 10 + 5 + 15 = 30
- B-D-F = 10 + 5 + 6 = 21
- C-D-E = 7 + 5 + 15 = 27
- C-D-F = 7 + 5 + 6 = 18.

Thus, B-D-E is the largest number (and therefore the longest duration). So it is the Critical Path.

One question that often comes up is "What happens if there is more than path which is the longest?" The answer is that all of them will be critical paths. Do note that all the durations will be the same value however. Obviously if there is variation, only the largest is the critical path. Can a project in the real world have more than one critical path? Yes, of course. We call those types of projects by several names. Failures, suicide projects, career suicide, and March to Hell are just some of the names. The more critical paths, the more work the project will take to keep it on track, and the more likely the project is to fail. But project managers who enjoy fear are perfectly within their rights to create multiple critical path projects. The rest of us will just learn to cry frequently and quake in our boots.

Step 3 - Calculate the early dates for the Critical Path items

Now that we know which path is the Critical Path (or which paths are the Critical Paths), we can begin the calculation of the schedule for that path. We call this set of calculations the forward pass. (For the sake of brevity, I'm going to stop saying path or paths. I'll just say path from now on. Please remember however, that there can be more than one Critical Path.)

Before we begin the actual forward pass, let's talk about how you calculate the two dates within the task box that we care about on this pass: the "Early Start" date and the "Early Finish" date.

Calculating Early Start dates

The "Early Start" date for all tasks is calculated by adding 1 to the "Early Finish" date of the latest predecessor. Let me explain why and then I'll clarify what I mean by latest predecessor. Remember when we discussed the format of the task node? If not look at Figure 1. You'll notice that the Start side of the task is as of start of day. The

Finish side is as of end of day. So let's say that I have task A running from start of day 1 to end of day 5, in other words say June 1 to June 5. All of June 5th has been used (let's say it's 11:59.59 pm). The next thing I do will be only 2 seconds later. That still means it will be 12:00.01 am on June 6, i.e. day 6 before we can begin the next task (say task D). Thus end of day on the "Early Finish" date of first task plus 1 gives us the start of day for the "Early Start" of the next task.

The next thing I need to explain is the concept of latest predecessor. Again, it is really simple. If I have multiple tasks, which must be completed before this task, I cannot begin this task until all predecessors have finished. Or to put it a slightly different way, until the last predecessor is finished (which implies the rest are also finished). In CPM terminology, we need to look for convergence. That is multiple paths coming together. A good example of this is Task D in Figure 2. All three paths, Tasks A, B, and C, must be completed before Task D can be started. Using our example (i.e. Figure 2), Task A runs from day 1 to day 5, Task B from day 1 to day 10, and Task C from day 1 to day 7. If Task D had only Task A in front of it, it could be started on Day 6 (= day 5 above + 1). If Task D had only Task B in front of it, it could be started on Day 11 (= day 10 above + 1). If Task D had only Task C in front of it, it could be started on Day 8 (= day 7 above + 1). However, that isn't the case. It has Task A, B, and C in front of it. Therefore, it can't be started until all three have been completed. This occurs on the last day for each of the paths plus one. In this case, Task B is the longest path (i.e. last day work is completed) so Task D can be started on Day 10 + 1 or Day 11.

Calculating Early Finish dates

"Early Finish" date is calculated almost as easily. However, there is one tricky item. To calculate "Early Finish" add the duration to the "Early Start" date and then subtract one. It's this last step (the subtract one) that seems to confuse new practitioners and be forgotten by both newbies and old hands. Remember that the Early Start is as of the beginning of the day, and Early Finish is as of the end of the day. Because the period is inclusive (that is it includes both Early Start and Early Finish), you need to subtract one. You may have heard the term "the Programmers' Calculator". Programmers often deal with series and calculations that either include the two numbers at the end, or exclude one or more of the two numbers at the end. This is why programmers tend to add and subtract using their fingers. Trying to remember if a sequence is inclusive or exclusive (and thus subtract one or not) leads to errors. So for programmers, it's easier to just count on the fingers. However, when doing CPM, it is always inclusive and therefore needs to account for the 1. This means that project managers get to look a little more professional when calculating. It's just not as much fun.

Performing the forward pass

Now that we know how to calculate the "Early Start" and "Early Finish" dates, we can begin calculating those dates for the Critical Path. We call this process the forward pass.

There are two basic techniques you will find being recommended. The first recommends that you start the first task in the path with a one as the "Early Start". And then calculate from there. There are two disadvantages with this technique. The first disadvantage is that it requires you to use two different rules for calculating "Early Start". One for the first task and one for every subsequent task.

The second disadvantage is that this technique is based on the project starting on day 1. In a complex project, this is not always the case.

For these reasons, I recommend that you use a modified technique. This is what I am going to discuss going forward.

For this technique, we are going to begin by assigning an "Early Finish" of zero to the Start milestone/terminus. What this is saying is that we will start as of the end of day on the 0th day of the project. This allows us to then calculate all tasks using the same rules.

Take the first task on the Critical Path and calculate the "Early Start" and "Early Finish" date using the rules above. So Early Start is the previous task's Early Finish plus one. And the Early Finish is the Early Start plus the duration less one. Now go to the next task on the Critical Path, and do the same thing. Keep doing this for each task on the Critical Path. Eventually you will hit the Finish milestone/terminus. For the Finish milestone/terminus, you simply calculate and record the "Early Start". This is the approximate date of the first entire day your project will be complete. You'll use this date in the backward pass for non-critical paths. We'll talk about those later.

If you have more than one Critical Path, pick one and calculate it. Then pick another and calculate it. Continue until you've finish calculating all the Critical Paths. The order you pick them in doesn't matter. Just be sure you don't pick a non-critical path. We'll deal with those in a moment.

One point that may come up as you are doing this is the concept of the milestone. Milestones are tasks with zero duration. They aren't really a task at all. What they are is a state marker. In other words, they are saying that we've achieved a certain state in the project. This typically means we've completed a certain group of tasks. The calculation of "Early Start" and "Early Finish" dates for milestones is exactly the same as shown above. The problem is that we end up with logical nonsense. We finish the day before we started. Time unfortunately can be a bit of a wibbly-wobbly-timey-wimey thing. Remember our discussion on start of day and end of day and subtracting one? That's the reason for the presentation being illogical. The start of today is the end of yesterday. So a task that occurs in the milliseconds between will always show finishing after it starts. To avoid having to explain this (and the resulting glazed eyes of management), we either don't show the dates at all (most commonly) or show only the "Early Start" date for milestones.

Step 4 - Record the late dates and the floats for the Critical Path items

You have now performed the forward pass, and calculated the "Early Start" and "Early Finish" dates. It's now time to calculate the "Late Start" and "Late Finish" dates. We call this the backward pass. The trick is with the Critical Path, this is simply a stub. We don't calculate anything. We simply copy.

Copy your dates from the forward pass line (at the top of the node) into the backward pass line (at the bottom). Make your "Late Start" date the same value as your "Early Start" date. Make your "Late Finish" date the same as your "Early Finish" date. The float is always zero. Do this for each of the Critical Path tasks.

You're now finished with the Critical Path(s). You also know how long the project will take. Of course, this date doesn't allow for non-work time, or delays due to staff availability, or sleep, or any other biological or rational function. But it is the minimum time you could complete the project. If you were working with an unlimited number of robots, of course. Which is all we are looking for at this point.

Step 5 - Calculate the early dates for each of the non-critical path items

Now that we've finished the Critical Path(s), it's time to begin working with the non-critical path tasks. Begin by picking one of the non-critical paths. It doesn't matter which one. However, I suggest you pick the simplest ones first. Avoid any with multiple predecessors if possible. A predecessor on the Critical Path is fine, but non-critical path predecessors can be a problem. We will cover all the tasks eventually but it's easier to do the simple ones first.

For each task in your chosen path, calculate the "Early Start" and "Early Finish" dates as described above in Step 3 for the Critical Path items. There is no difference in the rules. There is one thing to watch out for though – predecessors. If you have a Critical Path item as the predecessor, you can typically use that as the latest task. However, if you have only non-critical-path predecessors, you need to have calculated the "Early Finish" dates for each of the predecessors before you can calculate the "Early Start" for this task. That means you'll have to stop calculating this path and pick another. That's why you want to start with the simplest non-critical paths first. It prevents frustration. When you've finished go on to the next. When you finish the path, pick the next simplest and calculate those tasks. When you've calculated all the "Early Start" and "Early Finish" dates, you're done.

Step 6 - Calculate the late dates for each of the non-critical path items

Now that we've done the forward pass for each of the tasks, we're going to do the backward pass for the non-critical items. In essence, this is the reverse of the forward pass. Instead of worrying about convergence and multiple predecessors, we have to worry about divergence and multiple successors. Instead of performing the simple calculation of the "Early Start" date first and then the more complex "Early Finish" date based on that, we're going to calculate the "Late Finish" date first, and then perform the more complex "Late Start" calculation based on the result.

Calculating Late Finish dates

To calculate the "Early Start" date, we took the "Early Finish" date of the last predecessor and added one. We're going to work the other way for "Late Finish". We take the "Late Start" of the earliest successor, subtract one and we have the "Late Finish" of this task.

The concept of the earliest successor is exactly the reverse of the latest predecessor concept we have already encountered. Effectively, I need to finish this task before any of its successors can be started. Or looking at the situation from the same direction as we are working, you need to finish this task before the earliest date of any of the succeeding tasks' latest start dates. To illustrate let's consider the situation where I have two tasks that follow this task. Task E that needs to start by the start of the sixteenth day at the very latest. Task F needs to start no later than the start of the twenty-fifth day. This task (D), therefore, needs to finish no later than the end of the fifteenth day (Task E minus 1).

Calculating Late Start dates

Now that we know the "Late Finish" date, we can easily calculate the "Late Start" date for our task. The "Late Start" date is simply the "Late Finish" date plus one. Why the plus one? For exactly the same reason that we've subtracted one during the forward pass. Because the dates are inclusive.

If you are mathematical at all, the easy way to understand this is with a formula:

$$SD + D - 1 = FD$$

or if you prefer:

$$SD = FD - D + 1$$

where SD is the start date, FD is the finish date, and D is the duration. Both formulas are the same. If you have one, you can use basic algebra to derive the other. Which one you use depends on which value you have. On a forward pass, we have the start date so we use the first version of the formula. On a backward pass, we have the finish date, so we use the second version.

Performing the backward pass

Now that we know how to perform the calculations, we can perform the actual backward pass. Again, you begin by selecting the non-critical path. This time, however, you're going to be starting at the Finish milestone/terminus and working backwards. Last time it was beginning with the Start milestone/terminus and working forward. So start by looking for the simplest path. In other words, you look for a path with either no divergences or where there is only the Critical Path as an alternative successor.

Start at either the Finish milestone/terminus or where the path re-connects with the Critical Path. Then use the earliest "Late Start" dates of its successors to calculate the "Late Finish" for this task. This is why you want to record the "Early Start" date of the Finish milestone/terminus. In the same manner as you used the zero for the Start milestone/terminus as the predecessor, you'll use the "Start Date" of the Finish milestone/terminus to calculate the "Late Finish" of any non-critical paths tasks that end with the Finish milestone/terminus.

Once you have the "Late Finish" date, calculate the "Late Start" date. Work your way backwards through each task in the path. Once you finish each path, choose the next path. Work your way through that path. Keep repeating until all tasks have a "Late Start" and "Late Finish" date.

Step 7 - Calculate the float(s) for each of the non-critical path items

Now that we have calculated all the dates for all the tasks, the only thing left is to determine how much maneuver room we have. We call that maneuver room a task's float. Float applies only to non-critical tasks. Critical Path tasks have no float, that's why we call them critical, and why we recorded them as zero float during step 4, record the late dates and the floats for the Critical Path items. For non-critical path activities, there are two types of float.

The first is total float. It is how long the task can be delayed without damaging the project delivery date. It is shown on the bottom (or backward pass) line of the task node. The calculation is simple. Subtract the "Early Start" date from the "Late Start" date. The result will be the "Total Float" for the task. Alternatively, you could subtract the "Early Finish" date from the "Late Finish" date. Mathematically, both methods are identical. Record this amount on the bottom line. Go through all non-critical tasks and calculate them all. Notice that you no longer need to worry about the order you select the tasks in. Or the path they are in. As long as you do the calculation for all the non-critical tasks.

The second type of float is the free float. This float is amount of time that a task can be delayed without affecting the next task in the path. It is, of course, part of the total float for the task. In the early versions of CPM, there was no free float calculation, so the position to record the free float on the diagram varies from individual to individual.

Most record either above the node or above the line connecting the node and the successor. Personally, I prefer between but feel free to make your own choice. Most authors suggest only recording the value when it is non-zero.

Free float for a task can only occur when the succeeding task is delayed from starting because a third task has to finish first. For example, if Task A takes 5 days, Task B takes 10 days, and Task C takes seven days, then Task D which depends on all three cannot be started before day 11. Therefore Task A can be five days late and Task C can be three days late without affecting Task D's starting on time. The five and three represent the free float for Task A and Task C respectively. The actual calculation is "Early Start" of the successor minus "Early Finish" of this task minus one. Notice three things. First, the one in the calculation is the result of our inclusive dates (as usual). Second, that if the successor task was the Finish milestone/terminus, then the Free Float will equal the Total Float. And third, that only convergences (i.e. multiple predecessors) can create a free float situation. Note also that strictly speaking Free Float is associated with the path rather than the activity. Typically, we refer to the task's Free Float as the minimum of all the Free Floats from that activity.

When you've finished calculating the Total Float for all the non-critical tasks and the Free Float where they exist, you have completed the Critical Path Method calculations. You're now ready to perform the analysis and adjustment steps. In other words, start making the project plan achievable in the real world.

Putting it all together - an example

If we take the example presented in Figure 2 and perform the calculations, we get the following information:

Task	Early Start	Dur-ation	Early Finish	Late Finish	Late Start	Total Float
Start	0		0	0	0	
A	1 (0+1)	5	5 (1+5-1)	10 (11-1)	6 (10-5+1)	5
B	1 (0+1)	10	10 (1+10-1)	10 (CP)	1 (CP)	0
C	1 (0+1)	7	7 (1+7-1)	10 (11-1)	4 (10-7+1)	3
D	11 (10+1)	5	15 (11+5-1)	11 (CP)	15 (CP)	0
E	16 (15+1)	15	30 (16+15-1)	30 (CP)	16 (CP)	0
F	16 (15+1)	6	21 (16+6-1)	30 (31-1)	25 (30-6+1)	9
End	31		31	31	31	

Critical Path Analysis and beyond

Of course, calculating dates is not the full extent of the Critical Path Method. It is the full extent of this book however, so I'm not going to spend a great deal of time reviewing the remaining portions of the method. I will however, highlight some of the elements that you can expect to appear during your PMP® exam.

The first thing to be done is referred to, in CPM, as Critical Path Analysis. In the Guide to the PMBOK®, it is referred to as Schedule Network Analysis although elements are discussed under several other headings. The process is to look at the nature of the schedule, focusing on the critical path, in order to determine how to manipulate it. This includes the identification of parallel paths (divergence and convergence), resource reviews, and several other elements.

As a result of that analysis, you may identify that some portions of the plan are unrealistic. For example, you may be over-allocating your personnel. To correct this there are four alternatives you can use. They are:

1. Resource Optimization Techniques:
    a. Resource Leveling
    b. Resource Smoothing
2. Schedule Compression:
    a. Crashing
    b. Fast Tracking.

Both Resource Leveling and Resource Smoothing reduce workloads by extending the time required for specific activities. So for example, one individual might have four tasks to be completed in parallel. Each takes eight hours. Only one is critical. Uncorrected the individual would be scheduled for thirty-two hours of work in one day. Not exactly possible. To fix that situation, each of the four tasks is spread to take four days. Now the individual is working two hours per day on each task or a total of eight hours per day. Still not realistic but good enough for illustration purposes. With Resource Leveling, all tasks are subject to this spreading effect. No activity is sacrosanct. In Resource Smoothing, anything on the critical path is considered sacrosanct. Only tasks that aren't on the critical path are leveled. The result, of course, is that we may achieve nothing. In the example I gave, we still end up expecting thirty-two hours of work in a twenty-four hour period. Great performance if you could get it.

Resource Leveling and Smoothing leave the relationships between resources and tasks alone. They simply extend the period to achieve their goals. Crashing and Fast Tracking, however, are performed by changing the relationships between activities and resources. Crashing is throwing resources at the problem. You scheduled one person to work thirty-two hours in one day? Give the work to four people. Now they're only working eight hours. Or so the theory goes. Fast Tracking is performed by changing the relationship between tasks. Activities that were sequential are changed to become parallel. Of course, both of these techniques have downsides. Those downsides and limitations may eliminate any savings.

In addition to design and analysis, CPM also includes management and monitoring. Identification of the Critical Path tasks allows the project manager to focus on those tasks that are most likely to affect the final delivery of the project. And correspondingly less time on non-critical tasks that are unlikely to affect the final delivery.

Finally, in the sixty odd years since the development of CPM there have been many advances. Critical Chain Method is the most significant so far. CCM is an extension to CPM that helps deal with several problems related to CPM, estimation variation, and probabilities. More specifically, CPM was designed to work with the most likely dates. But the PERT method established that the actual duration will vary in a range around the most likely. To be more precise it varies around the expected value or the mean (aka the average) which is typically close to the most likely. Most project managers deal with the problem either by hoping that the net variation will be near to zero or by using a value closer to the pessimistic. The latter choice is usually performed with a simple 'fudge factor'.

Unfortunately, one can't simply add statistical quantities together. The mathematics of stats is different from that of simple arithmetic. Adding the most likely amounts of each of the activities is not statistically justifiable (i.e. it gives a wrong answer).

In addition, there is the issue of basic human nature. Specifically, there is a tendency for the time required to match the time available. If we believe we have more time available, we'll work slower. We've all seen tasks that should have taken five minutes that end up killing a morning (or more). Telling people they have more than the minimum results in a bias to the high end of the estimation range.

To overcome these issues, Critical Chain performs the Critical Path Method using the minimum values (i.e. the optimistic values). It then creates tasks called buffers equal to the statistical sum of the variances. These buffers are placed at the end of the critical path and at various other points in the activity network. The critical path is then recalculated. The result is a statistically justifiable critical path. As the project progresses, the actual amounts used in excess of the optimistic are drawn from the appropriate buffer amount. You can then manage the project by monitoring the amount of buffer burned and the amount of buffer remaining in comparison to what you expected to burn.

The actual process of Critical Chain Method is, of course, more complex than the admittedly simple explanation above. Since this isn't a book on Critical Chain Method, I'm not going to discuss the method in any more detail. If you are interested, I recommend to begin with the information in the PMBOK® GUIDE and then read one of the many books and articles on using the Critical Chain Method.

Doing it wrong

Earlier, I mentioned that there were several methods of doing CPM correctly. What I was referring to primarily was two choices. Whether you use a seven-step or five-step method is one. Whether you give dates to the start and finish terminators being the other choice. I've chosen to use the seven-step with dates on the end nodes because it is the easiest to understand. The process is actually the same regardless of your choice.

However, in the past sixty years (or more) a number of variations have occurred which aren't correct. They lead to incorrect answers, to flawed results, extra work, or increase the probability of a wrong answer. It is important that you be aware of them. You definitely don't want to follow the advice of someone who suggests one of these methods. I've been around a long time, but even I haven't seen all the ways to mess up the process. However, I can give you two of the more common errors.

The first error is to skip identifying the critical path. Just begin at the beginning and calculate. The critical path will appear on its own. I can't begin to tell you how wrong this is. Well, I can begin but I can't possibly include all the reasons. The biggest is the fact that you are going to end up doing twice the work. This method requires you to calculate the backward pass for all activities rather than just copying for critical path activities. That means you have twice the opportunity for making a calculation mistake. In addition, if your project has multiple convergences and divergences, you are going to find yourself restarting and changing paths frequently. This increases the amount of work (and probability of mistakes) immensely.

The second error is using exclusive dates. By expressing the finish date as the start of the day, you are able to perform the calculations without using the plus or minus one adjustment. While there are many reasons to prefer this method, there is one overriding reason to dismiss this method. The answer is wrong on the PMP® exam. Yes, I know that is a ridiculous reason but ultimately, that's the biggest reason to not use it. Having said that, there actually is a reason beyond "PMI® said to". There are actually three different configurations possible. Inclusive, shows the start dates as of the beginning of the day, and the finish dates as of the end of the day. It always shows a one-day difference between finish of an activity and the start of the next. However, there are two possible exclusive methods. You can show both dates as of the beginning of the date shown. Or you can show both dates as of the end of day shown. Although they give a one time-period difference, there is no easy way to distinguish between them. If the period is one day, it may not matter. If the period shown is in weeks or months, that lack of clarity may be a serious issue. Choosing the one method that is easily distinguished eliminates misreading of the dates. That's why the inclusive method was selected, even though it represents extra work when calculating.

Chapter 3: The Exercises

The following CPM exercises are of three types.

The first type of exercise involves creating a CPM chart from a PDM chart. In those exercises, you will find a PDM chart. Your job will be to calculate all of the information and turn the PDM charts into CPM charts. You can see the previous chapter if you need help. Each of the charts is followed on the next page by the answer. The answer should be sufficient to determine the method used to calculate it, but if not, a step-by-step solution is also shown. The exercises start with very simple questions. The questions become progressively harder. This will help you to develop some confidence in your abilities. The exercises will, in fact, become much harder than the examples used in the PMP® exam. After all, there is no time limit on these exercises unlike the exam. All examples use the same building blocks so if you can do the more complex exercises, you will find the simpler exercises easy.

The second type of exercises will be focused on interpretation. A situation will be presented and you will be expected to determine the answer. This will require you to draw and complete a PDM chart, and then perform the CPM calculations.

The first two types of exercises are mixed together and the difficulty is clearly marked. While you don't have to perform all the exercises, I highly recommend you complete at least some from each difficulty level.

The third type of exercise is separated and follows the other two types (questions 41 to 50). These are interpretive questions. The information you are given and the information you need to answer will vary. Information will be presented in a PDM chart with values missing. You will then need to calculate a set of values from the ultimate chart. I suggest that you may want to test yourself by being efficient and calculating only those numbers that you actually need for the answer. Effectively, you are testing yourself to go beyond rote and understand how the numbers relate. Once you've answered the question efficiently, only then should you calculate all the missing information (PDM chart, CPM calculations etc.). The answers are presented as a complete CPM chart including all the missing information.

While I suggest that you complete all the exercises, you really only need to complete enough exercises from each group to be comfortable that you know the process. However, you do need to treat the three groups as independent. In other words, you need to complete some questions from set one and some from set two and some from set three. In addition, as I mentioned, group one and two questions vary in difficulty. I suggest that you need to answer some from each difficulty group.

In all cases, the answer will immediately follow the question. I recommend that you have a pad of paper and pencil as you answer questions. Answer the question and then compare your answer to the given answer. Do not look at the answer before you have at least attempted to answer the question. This is especially important with the electronic version of this book. In that version, it is possible to see both the question and the answer at the same time. You will need to ensure that does not happen.

Good luck with the exercises. And good luck with the exam.

Question 1 - Calculate the values on the CPM chart (Very Simple)

Scenario:

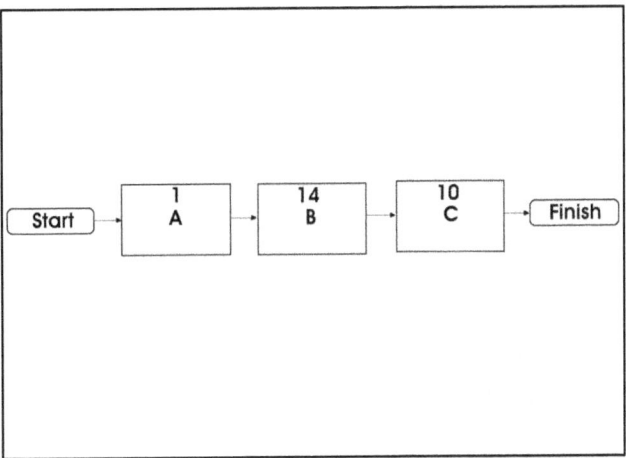

Question:

Given the above information, identify the critical path, calculate all the dates, and calculate the floats.

Answer:

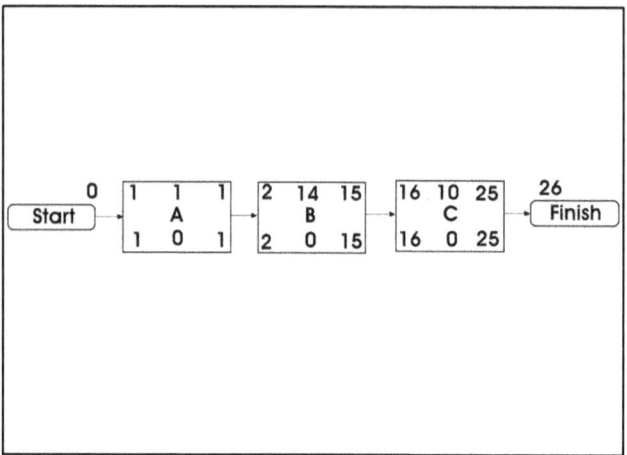

Since there are only three tasks A, B, and C performed sequentially, there is only one path. The critical path is therefore A-B-C.

Question 2 - Develop the CPM chart (Very Simple)

Question:

Your project consists of three tasks, A, B, and C. All three are done sequentially. Task A is expected to take 5 days, task B is expected to take 1 day and task C is expected to take 10 days. Diagram the project using PDM, then prepare the CPM chart, identify the critical path, calculate all the dates, and calculate the floats.

Answer:

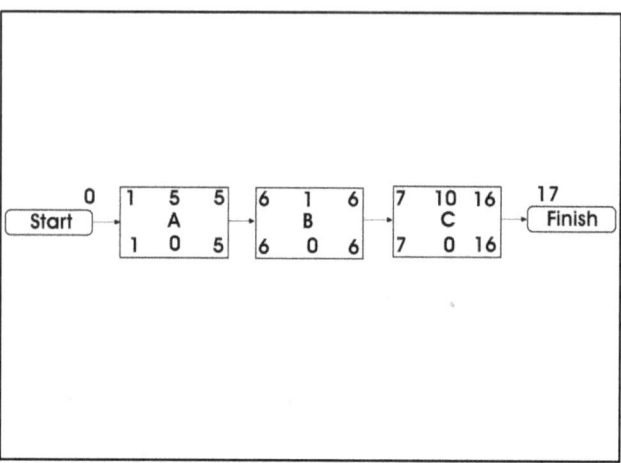

Process:

1. Design chart, placing letter or process name in the middle, and the duration on the top row in the middle.

2. Since there are only three tasks A, B, and C performed sequentially, there is only one path. The critical path is therefore A-B-C.

3. Calculate the forward pass for the critical path A-B-C. This goes on the upper row.

4. Perform the backward pass on the critical path. To do this, copy the Early Start and Early Finish from the upper row into the Late Start and Late Finish. The Late Start and Late Finish go into the bottom row.

5. Calculate the floats for the critical path. This is always zero and is placed in the middle position of the bottom row. The free float is also zero and therefore is ignored.

Question 3 - Calculate the values on the CPM chart (Very Simple)

Scenario:

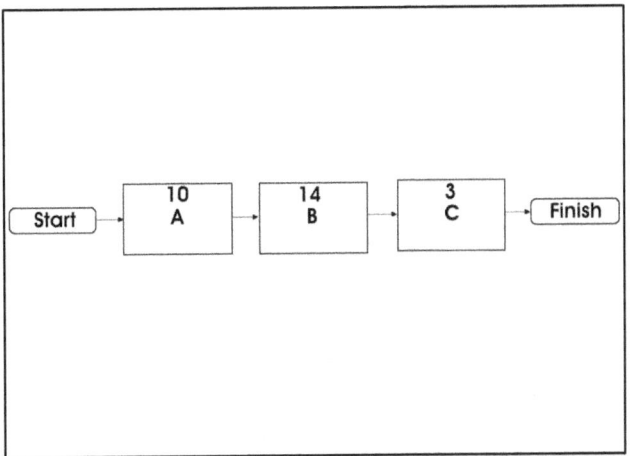

Question:

Given the above information, identify the critical path, calculate all the dates, and calculate the floats.

Answer:

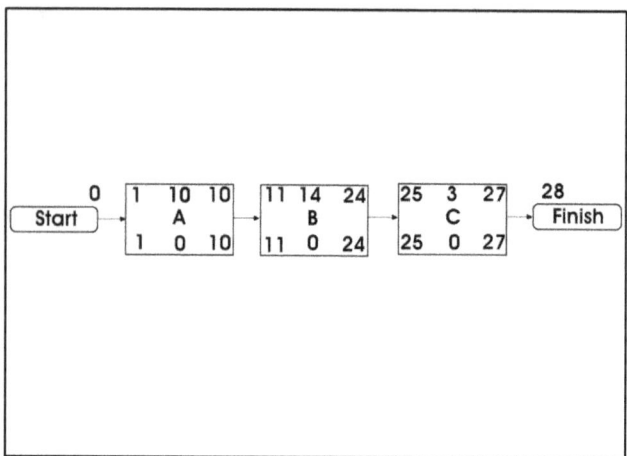

Since there are only three tasks A, B, and C performed sequentially, there is only one path. The critical path is therefore A-B-C.

Question 4 - Develop the CPM chart (Very Simple)

Question:

Your project consists of three tasks, A, B, and C. They are done sequentially. Task A is expected to take 12 days, task B is expected to take 3 days and task C is expected to take 2 days. Diagram the project using PDM, then prepare the CPM chart, identify the critical path, calculate all the dates, and calculate the floats.

Answer:

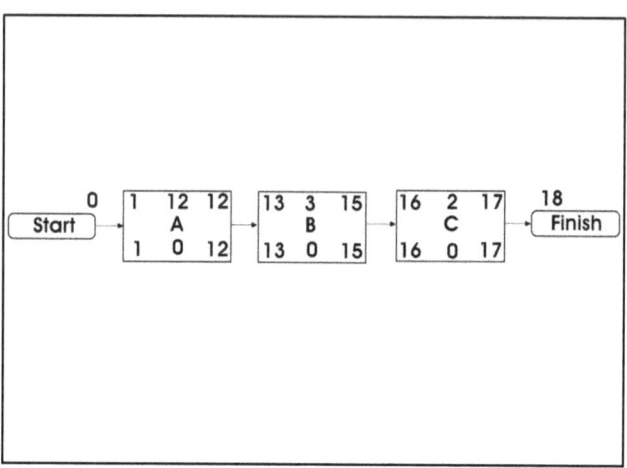

Process:

1. Design chart, placing letter or process name in the middle, and the duration on the top row in the middle.

2. Since there are only three tasks A, B, and C performed sequentially, there is only one path. The critical path is therefore A-B-C.

3. Calculate the forward pass for the critical path A-B-C. This goes on the upper row.

4. Perform the backward pass on the critical path. To do this, copy the Early Start and Early Finish from the upper row into the Late Start and Late Finish. The Late Start and Late Finish go into the bottom row.

5. Calculate the floats for the critical path. This is always zero and is placed in the middle position of the bottom row. The free float is also zero and therefore is ignored.

Scenario:

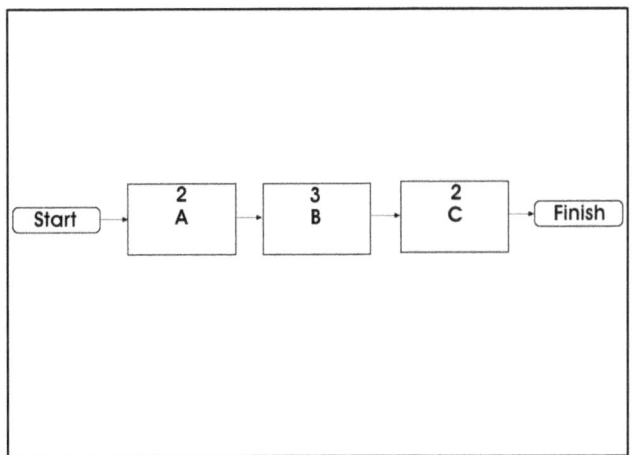

Question:

Given the above information, identify the critical path, calculate all the dates, and calculate the floats.

Answer:

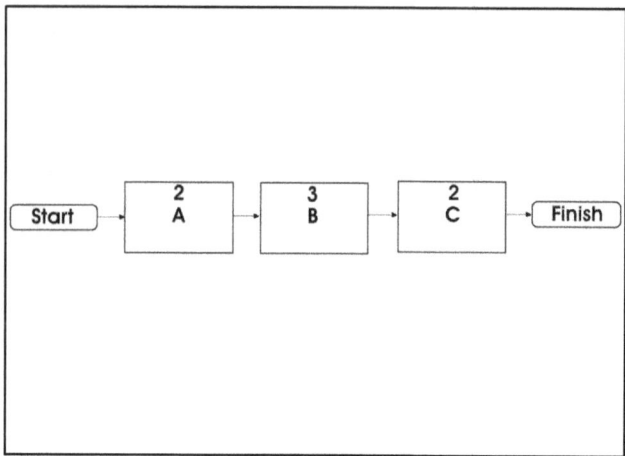

Since there are only three tasks A, B, and C performed sequentially, there is only one path. The critical path is therefore A-B-C.

Question 6 - Develop the CPM chart (Very Simple)

Question:

Your project consists of three tasks, A, B, and C. All three are done sequentially. Task A is expected to take 7 days, task B is expected to take 3 days and task C is expected to take 4 days. Diagram the project using PDM, then prepare the CPM chart, identify the critical path, calculate all the dates, and calculate the floats.

Answer:

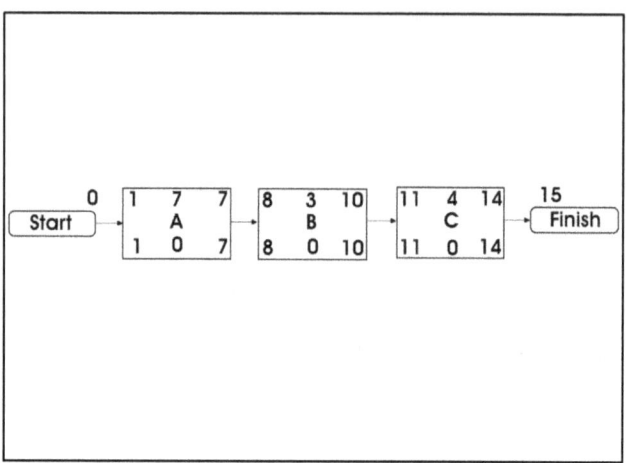

Process:

1. Design chart, placing letter or process name in the middle, and the duration on the top row in the middle.

2. Since there are only three tasks A, B, and C performed sequentially, there is only one path. The critical path is therefore A-B-C.

3. Calculate the forward pass for the critical path A-B-C. This goes on the upper row.

4. Perform the backward pass on the critical path. To do this, copy the Early Start and Early Finish from the upper row into the Late Start and Late Finish. The Late Start and Late Finish go into the bottom row.

5. Calculate the floats for the critical path. This is always zero and is placed in the middle position of the bottom row. The free float is also zero and therefore is ignored.

Question 7 - Calculate the values on the CPM chart (Very Simple)

Scenario:

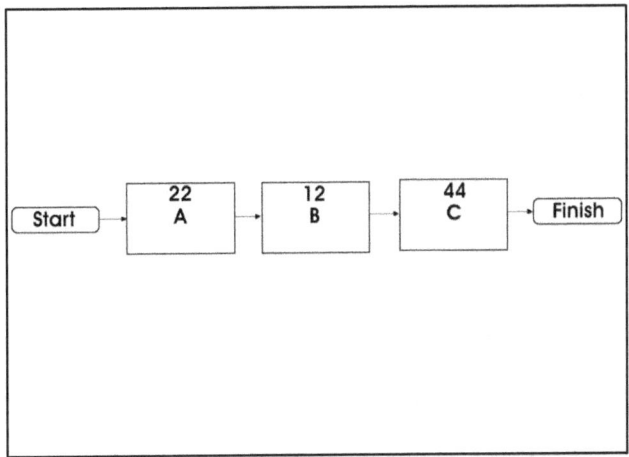

Question:

Given the above information, identify the critical path, calculate all the dates, and calculate the floats.

Answer:

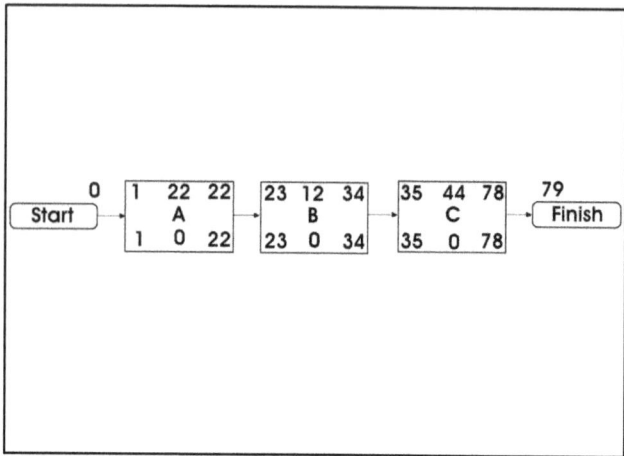

Since there are only three tasks A, B, and C performed sequentially, there is only one path. The critical path is therefore A-B-C.

Question 8 - Develop the CPM chart (Very Simple)

Question:

Your project consists of three tasks, A, B, and C. They are done sequentially. Task A is expected to take 73 days, task B is expected to take 12 days and task C is expected to take 2 days. Diagram the project using PDM, then prepare the CPM chart, identify the critical path, calculate all the dates, and calculate the floats.

Answer:

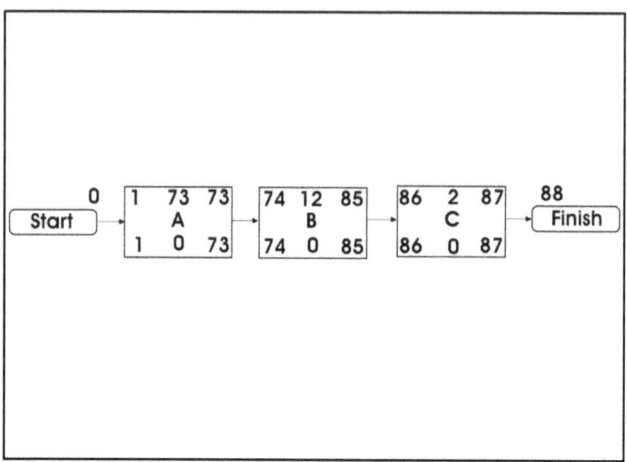

Process:

1. Design chart, placing letter or process name in the middle, and the duration on the top row in the middle.

2. Since there are only three tasks A, B, and C performed sequentially, there is only one path. The critical path is therefore A-B-C.

3. Calculate the forward pass for the critical path A-B-C. This goes on the upper row.

4. Perform the backward pass on the critical path. To do this, copy the Early Start and Early Finish from the upper row into the Late Start and Late Finish. The Late Start and Late Finish go into the bottom row.

5. Calculate the floats for the critical path. This is always zero and is placed in the middle position of the bottom row. The free float is also zero and therefore is ignored.

Question 9 - Calculate the values on the CPM chart (Very Simple)

Scenario:

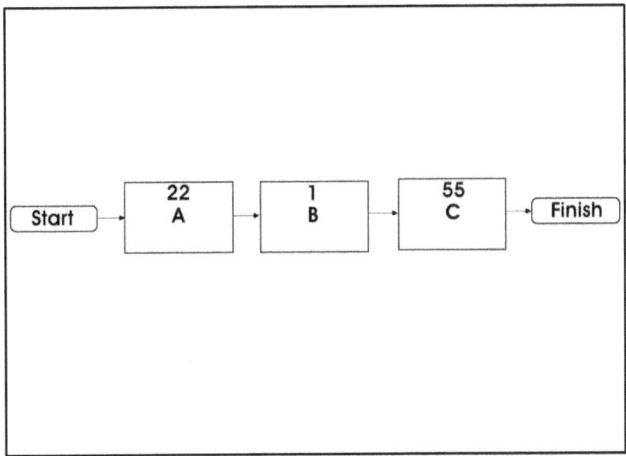

Question:

Given the above information, identify the critical path, calculate all the dates, and calculate the floats.

Answer:

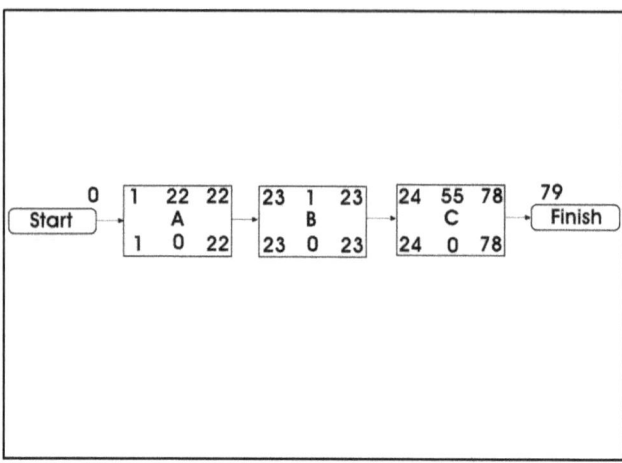

Since there are only three tasks A, B, and C performed sequentially, there is only one path. The critical path is therefore A-B-C.

Question 10 - Develop the CPM chart (Very Simple)

Question:

Your project consists of three tasks, A, B, and C. They are done sequentially. Task A is expected to take 1 days, task B is expected to take 105 days and task C is expected to take 55 days. Diagram the project using PDM, then prepare the CPM chart, identify the critical path, calculate all the dates, and calculate the floats.

Answer:

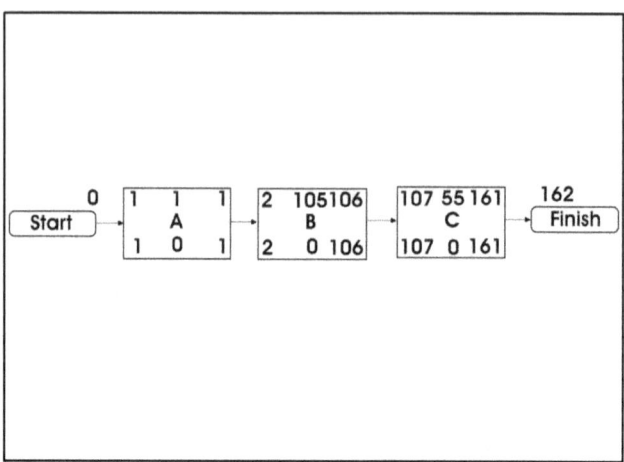

Process:

1. Design chart, placing letter or process name in the middle, and the duration on the top row in the middle.

2. Since there are only three tasks A, B, and C performed sequentially, there is only one path. The critical path is therefore A-B-C.

3. Calculate the forward pass for the critical path A-B-C. This goes on the upper row.

4. Perform the backward pass on the critical path. To do this, copy the Early Start and Early Finish from the upper row into the Late Start and Late Finish. The Late Start and Late Finish go into the bottom row.

5. Calculate the floats for the critical path. This is always zero and is placed in the middle position of the bottom row. The free float is also zero and therefore is ignored.

Question 11 - Calculate the values on the CPM chart (Simple)

Scenario:

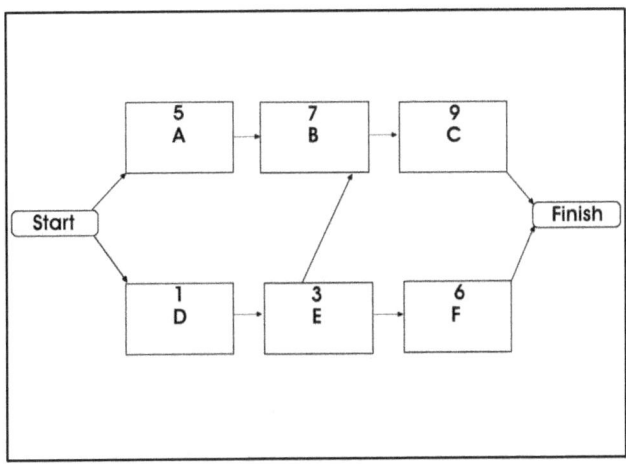

Question:

Given the above information, identify the critical path, calculate all the dates, and calculate the floats.

Answer:

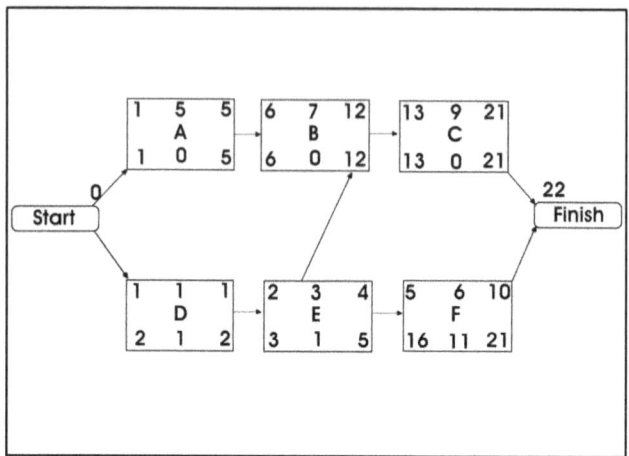

There are three paths A-B-C, D-E-B-C, and D-E-F. A-B-C has a total duration of 5+7+9=21. D-E-B-C has a total duration of 1+3+7+9=20. D-E-F has a total duration of 1+3+6=10. A-B-C is the longest so it is the critical path and is to be calculated first.

The path E-B has a free float of 6-4-1=1 day. The path E-F has a free float of 5-4-1=0 days. (E is generally considered to have a free float of the minimum of these two). F to finish has a free float equal to its total float by definition. All other free floats are 0.

Question 12 - Develop the CPM chart (Simple)

Question:

Your project consists of six tasks, A, B, C, D, E, and F. Task A is expected to take 6 days, task B is expected to take 7 days, task C is expected to take 9 days, task D is expected to take 4 days, task E is expected to take 3 days, and task F is expected to take 6 days. Task A and D begin concurrently. Task B is preceded by both task A and task E. Task C is preceded by task B and ends the path. Task D precedes task E. Task F is preceded by task E and is the last task. Diagram the project using PDM, then prepare the CPM chart, identify the critical path, calculate all the dates, and calculate the floats.

Answer:

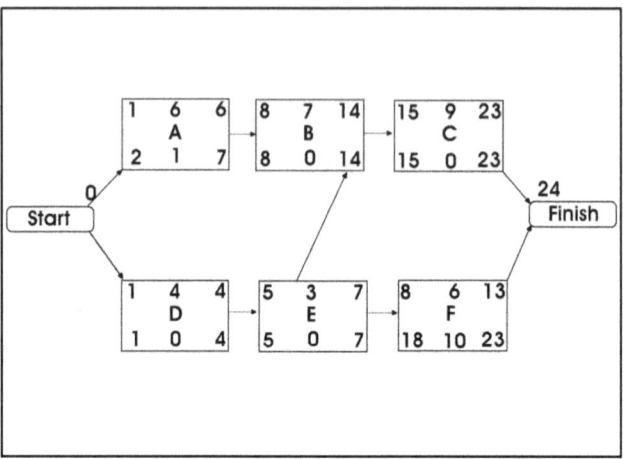

Process:

1. Design chart, placing letter or process name in the middle, and the duration on the top row in the middle.

2. There are three paths A-B-C, D-E-B-C, and D-E-F. A-B-C has a total duration of 6+7+9=22. D-E-B-C has a total duration of 4+3+7+9=23. D-E-F has a total duration of 4+3+6=13. D-E-B-C is the longest so it is the critical path and is to be calculated first.

3. Calculate the forward pass for the critical path D-E-B-C. This goes on the upper row of the critical tasks.

4. Perform the backward pass on the critical path. To do this, copy the Early Start and Early Finish from the upper row of the critical tasks into the Late Start and Late Finish. The Late Start and Late Finish go into the bottom row of the critical tasks.

5. Calculate the floats for the critical path. This is always zero and is placed in the middle position of the bottom row of the critical tasks. The free float is also zero and therefore is ignored.

6. Calculate the forward paths for the non-critical tasks: A and F. This is placed on the top row of those tasks.

7. Calculate the backward paths for the non-critical tasks: A and F. This is placed on the bottom row of those tasks.

8. Calculate the total float for tasks A and F. This is placed in the middle position of the bottom row of those tasks.

9. Calculate the free float for A-B (8-6-1=1). This is placed near the line on A. The free float for F-Finish is the same as the total float for F and doesn't have to be calculated again.

Question 13 - Calculate the values on the CPM chart (Simple)

Scenario:

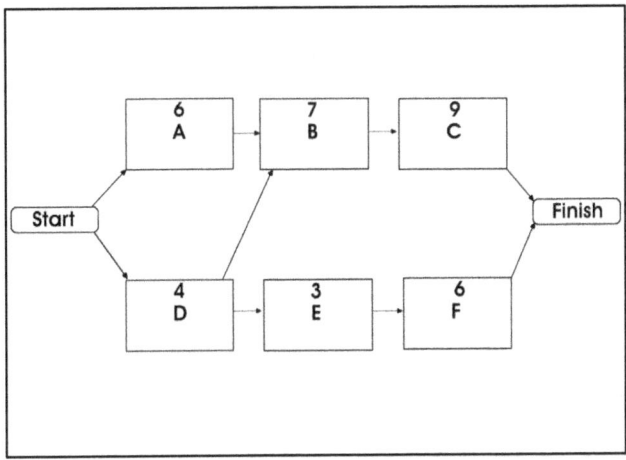

Question:

Given the above information, identify the critical path, calculate all the dates, and calculate the floats.

Answer:

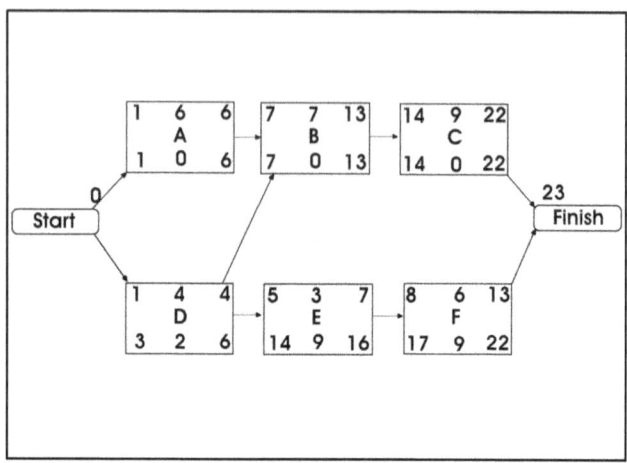

Notice that this and question 12 are almost identical. Only the predecessor to B has changed. The result is considerably different however. There are three paths A-B-C, D-B-C, and D-E-F. A-B-C has a total duration of 6+7+9=22. D-B-C has a total duration of 4+7+9=20. D-E-F has a total duration of 4+3+6=13. A-B-C is the longest so it is the critical path and is to be calculated first.

The path D-B has a free float of 7-4-1=2 while the path D-E has a free float of 5-4-1=0 (D is generally held to have a free float of the minimum of these two). The path F to finish has a free float equal to its total float by definition. All other free floats (i.e. E-F) are 0.

Question 14 - Develop the CPM chart (Simple)

Question:

Your project consists of six tasks, A, B, C, D, E, and F. Task A is expected to take 6 days, task B is expected to take 7 days, task C is expected to take 9 days, task D is expected to take 4 days, task E is expected to take 3 days, and task F is expected to take 6 days. Task A and D begin concurrently. Task B is preceded by task A. Task C is preceded by both task B and task E and ends the path. Task D precedes task E. Task F is preceded by task E and is the last task on the path. Diagram the project using PDM, then prepare the CPM chart, identify the critical path, calculate all the dates, and calculate the floats.

Answer:

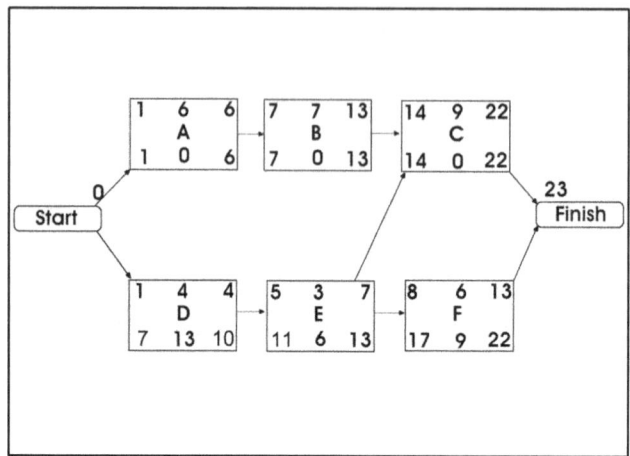

Process:

1. Design chart, placing letter or process name in the middle, and the duration on the top row in the middle.

2. There are three paths A-B-C, D-E-C, and D-E-F. A-B-C has a total duration of 6+7+9=22. D-E-C has a total duration of 4+3+9=16. D-E-F has a total duration of 4+3+6=13. A-B-C is the longest so it is the critical path and is to be calculated first.

3. Calculate the forward pass for the critical path A-B-C. This goes on the upper row of the critical tasks.

4. Perform the backward pass on the critical path. To do this, copy the Early Start and Early Finish from the upper row of the critical tasks into the Late Start and Late Finish. The Late Start and Late Finish go into the bottom row of the critical tasks.

5. Calculate the floats for the critical path. This is always zero and is placed in the middle position of the bottom row of the critical tasks. The free float is also zero and therefore is ignored.

6. Calculate the forward paths for the non-critical tasks: D, E, and F. This is placed on the top row of those tasks.

7. Calculate the backward paths for the non-critical tasks: D, E, and F. This is placed on the bottom row of those tasks.

8. Calculate the total float for tasks D, E, and F. This is placed in the middle position of the bottom row of those tasks.

9. Calculate the free float for E-C (14-7-1=6). This is placed near the line on E. Calculate the free float for E-F (8-7-1=0). (The free float for task E is generally considered to be the minimum of those paths). The free float for F-Finish is the same as the total float for F and doesn't have to be calculated again. All other free floats (i.e. D-E) are 0.

Question 15 - Calculate the values on the CPM chart (Simple)

Scenario:

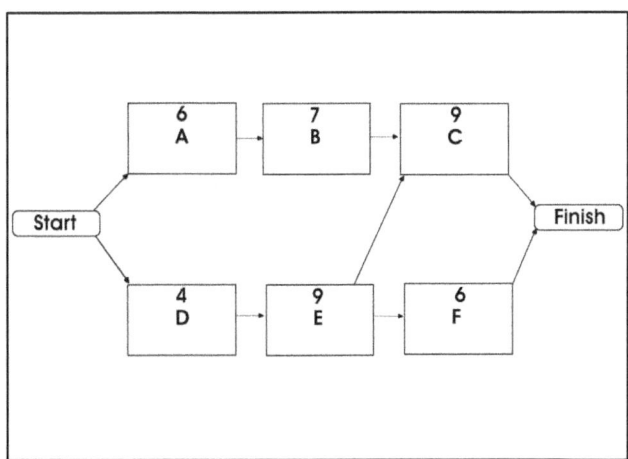

Question:

Given the above information, identify the critical path, calculate all the dates, and calculate the floats.

Answer:

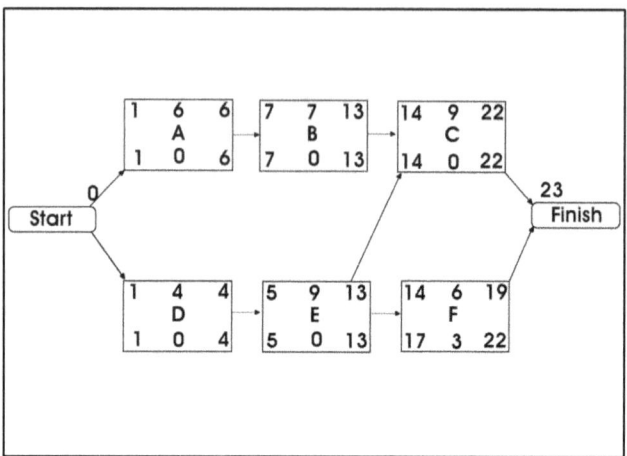

There are three paths A-B-C, D-E-C, and D-E-F. A-B-C has a total duration of 6+7+9=22. D-E-C has a total duration of 4+9+9=22. D-E-F has a total duration of 4+9+6=19. A-B-C and D-E-C are equivalent and both are the longest so they are both critical paths and to be calculated first.

The path F to finish has a free float equal to its total float by definition. F is the only non-critical activity.

Question 16 - Develop the CPM chart (Simple)

Question:

Your project consists of six tasks, A, B, C, D, E, and F. Task A is expected to take 10 days, task B is expected to take 8 days, task C is expected to take 9 days, task D is expected to take 4 days, task E is expected to take 10 days, and task F is expected to take 6 days. Task A and D begin concurrently. Task B is preceded by task A. Task C is preceded by both task B and task E and ends the path. Task D precedes task E. Task F is preceded by task E and is the last task on the path. Diagram the project using PDM, then prepare the CPM chart, identify the critical path, calculate all the dates, and calculate the floats.

Answer:

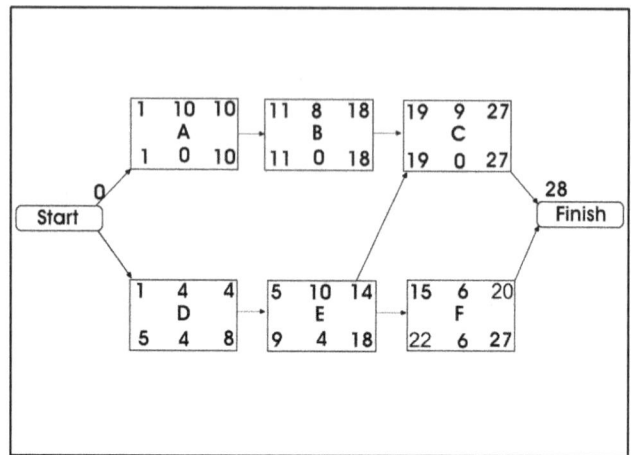

Process:

1. Design chart, placing letter or process name in the middle, and the duration on the top row in the middle.

2. There are three paths A-B-C, D-E-C, and D-E-F. A-B-C has a total duration of 10+8+9=27. D-E-C has a total duration of 4+10+9=23. D-E-F has a total duration of 4+10+6=20. A-B-C is the longest so it is the critical path and is to be calculated first.

3. Calculate the forward pass for the critical path A-B-C. This goes on the upper row of those critical tasks.

4. Perform the backward pass on the critical path. To do this, copy the Early Start and Early Finish from the upper row of the critical tasks into the Late Start and Late Finish. The Late Start and Late Finish go into the bottom row of the critical tasks.

5. Calculate the floats for the critical path. This is always zero and is placed in the middle position of the bottom row of the critical tasks. The free float is also zero and therefore is ignored.

6. Calculate the forward paths for the non-critical tasks: D, E, and F. This is placed on the top row of these non-critical tasks.

7. Calculate the backward paths for the non-critical tasks: D, E, and F. This is placed on the bottom row of these non-critical tasks.

8. Calculate the total float for tasks D, E, and F. This is placed in the middle position of the bottom row of these non-critical tasks.

9. Calculate the free float for E-C (19-14-1=4). This is placed near the line on E. The free float for F-Finish is the same as the total float for F and doesn't have to be calculated again. All other free floats (i.e. D-E, and E-F) are 0.

Question 17 - Calculate the values on the CPM chart (Simple)

Scenario:

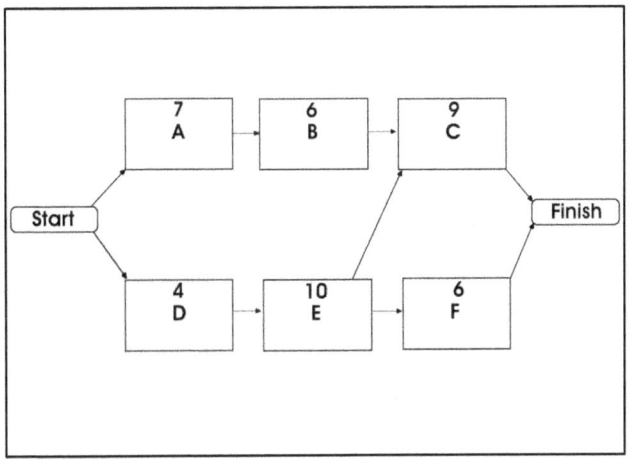

Question:

Given the above information, identify the critical path, calculate all the dates, and calculate the floats.

Answer:

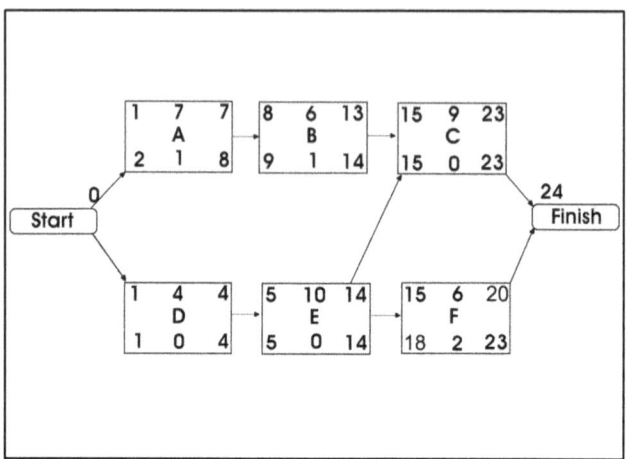

There are three paths A-B-C, D-E-C, and D-E-F. A-B-C has a total duration of 7+6+9=22. D-E-C has a total duration of 4+10+9=23. D-E-F has a total duration of 4+10+6=20. D-E-C is the longest so it is the critical path and to be calculated first.

The path F to finish has a free float equal to its total float by definition. B to C also has a free float equal to its total float by definition. Only path A to B needs to be calculated. It has a free float of (8-7-1=) 0.

Question 18 - Develop the CPM chart (Simple)

Question:

Your project consists of six tasks, A, B, C, D, E, and F. Task A is expected to take 3 days, task B is expected to take 6 days, task C is expected to take 9 days, task D is expected to take 2 days, task E is expected to take 7 days, and task F is expected to take 6 days. Task A and D begin concurrently. Task B is preceded by task A. Task C is preceded by both task B and task E and ends the path. Task D precedes task E. Task F is preceded by task E and is the last task on the path. Diagram the project using PDM, then prepare the CPM chart, identify the critical path, calculate all the dates, and calculate the floats.

Answer:

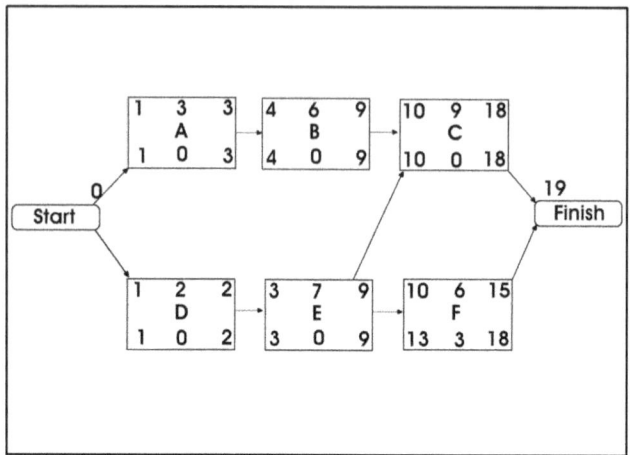

Process:

1. Design chart, placing letter or process name in the middle, and the duration on the top row in the middle.

2. There are three paths A-B-C, D-E-C, and D-E-F. A-B-C has a total duration of 3+6+9=18. D-E-C has a total duration of 2+7+9=18. D-E-F has a total duration of 2+7+6=15. A-B-C and D-E-C are equivalent and both are the longest so they are both critical paths and to be calculated first.

3. Calculate the forward pass for the critical path A-B-C. This goes on the upper row. Then calculate the forward pass for the critical path D-E-C. This goes on the upper row of D and E. Notice that C has already been calculated.

4. Perform the backward pass on the critical path. To do this, copy the Early Start and Early Finish from the upper row into the Late Start and Late Finish. The Late Start and Late Finish go into the bottom row of the critical tasks.

5. Calculate the floats for the critical path. This is always zero and is placed in the middle position of the bottom row of the critical tasks. The free float is also zero and therefore is ignored.

6. Calculate the forward paths for the non-critical task F. This is placed on the top row of task F.

7. Calculate the backward paths for the non-critical task F. This is placed on the bottom row of task F.

8. Calculate the total float for task F. This is placed in the middle position of the bottom row of task F.

9. The free float for F-Finish is the same as the total float for F and doesn't have to be calculated again.

Question 19 - Calculate the values on the CPM chart (Simple)

Scenario:

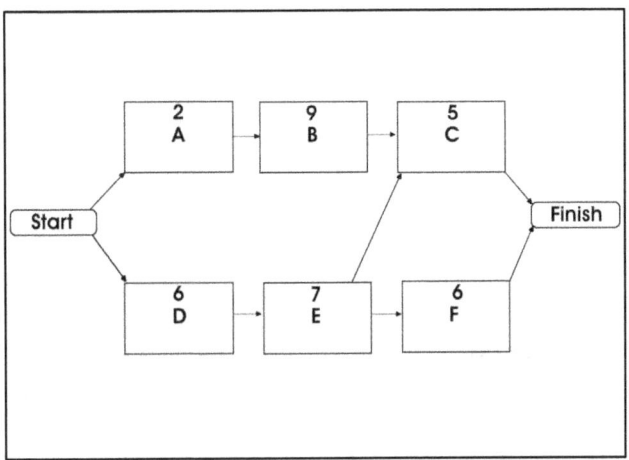

Question:

Given the above information, identify the critical path, calculate all the dates, and calculate the floats.

Answer:

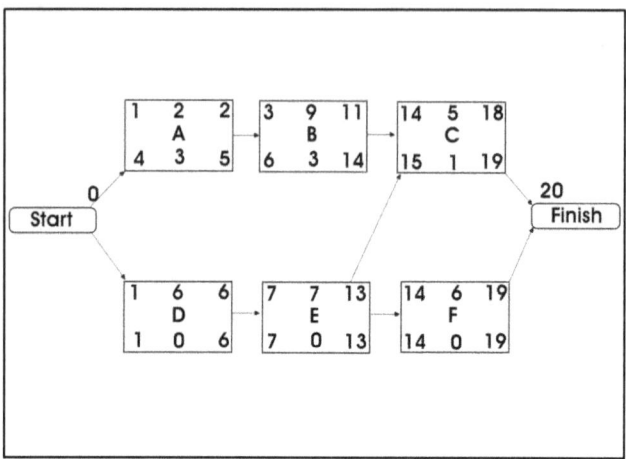

There are three paths A-B-C, D-E-C, and D-E-F. A-B-C has a total duration of 2+9+5=16. D-E-C has a total duration of 6+7+5=18. D-E-F has a total duration of 6+7+6=19. D-E-F is the longest so it is the critical path and to be calculated first.

The path C to finish has a free float equal to its total float by definition. B to C has a free float of (14-11-1=) 2. Path A to B has a free float of (3-2-1=) 0.

Question 20 - Develop the CPM chart (Simple)

Question:

Your project consists of six tasks, A, B, C, D, E, and F. Task A is expected to take 2 days, task B is expected to take 9 days, task C is expected to take 6 days, task D is expected to take 8 days, task E is expected to take 4 days, and task F is expected to take 6 days. Task A and D begin concurrently. Task B is preceded by task A. Task C is preceded by both task B and task E and ends the path. Task D precedes task E. Task F is preceded by task E and is the last task on the path. Diagram the project using PDM, then prepare the CPM chart, identify the critical path, calculate all the dates, and calculate the floats.

Answer:

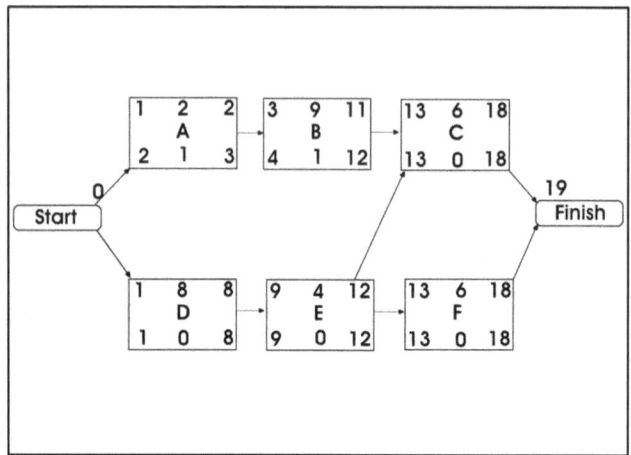

Process:

1. Design chart, placing letter or process name in the middle, and the duration on the top row in the middle.

2. There are three paths A-B-C, D-E-C, and D-E-F. A-B-C has a total duration of 2+9+6=17. D-E-C has a total duration of 8+4+6=18. D-E-F has a total duration of 8+4+6=18. D-E-C and D-E-F are equivalent and both are the longest so they are both critical paths and to be calculated first.

3. Calculate the forward pass for the critical path D-E-C. This goes on the upper row of critical tasks D, E, and C. Then calculate the forward pass for the critical path D-E-F. This is the same as saying calculate activity F since D and E have already been calculated. The result goes on the upper row of F.

4. Perform the backward pass on the critical paths. To do this, copy the Early Start and Early Finish from the upper row into the Late Start and Late Finish. The Late Start and Late Finish go into the bottom row of the critical tasks.

5. Calculate the floats for the critical paths. This is always zero and is placed in the middle position of the bottom row of the critical tasks. The free float is also zero and therefore is ignored.

6. Calculate the forward paths for the non-critical tasks A and B. This is placed on the top row of non-critical tasks A and B.

7. Calculate the backward paths for the non-critical tasks A and B. This is placed on the bottom row of tasks A and B.

8. Calculate the total float for tasks A and B. This is placed in the middle position of the bottom row of tasks A and B.

9. The free float for B-C is the same as the total float for B by definition and doesn't have to be calculated again. The free float for A to B is (3-2-1=) 0.

Question 21 - Calculate the values on the CPM chart (Hard)

Scenario:

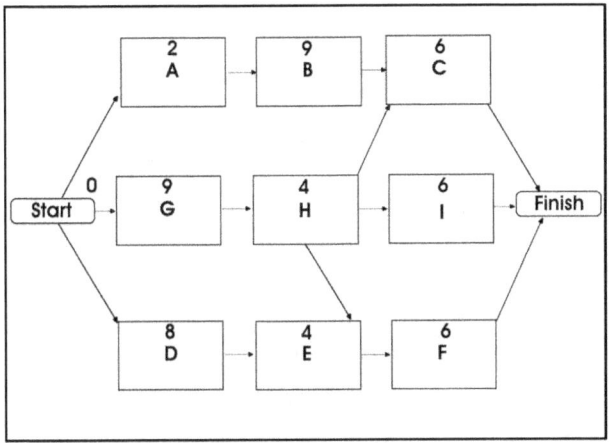

Question:

Given the above information, identify the critical path, calculate all the dates, and calculate the floats.

Answer:

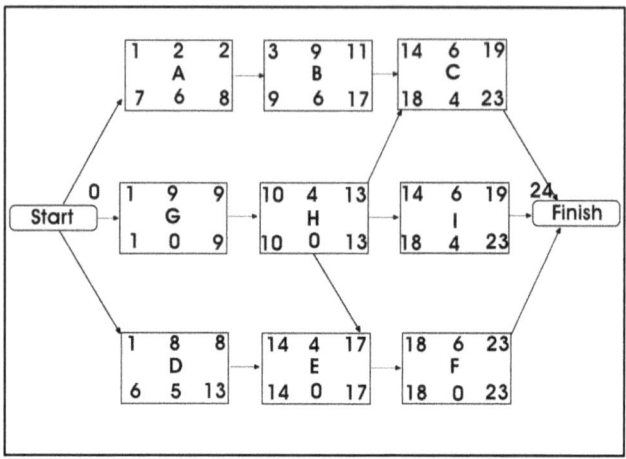

There are five paths A-B-C, D-E-F, G-H-I, G-H-C, and G-H-E-F. A-B-C has a total duration of 2+9+6=17. G-H-I has a total duration of 9+4+6=19. D-E-F has a total duration of 8+4+6=18. G-H-C has a total duration of 9+4+6=19. G-H-E-F has a total duration of 9+4+4+6=23. G-H-E-F is the longest so it is the critical path and to be calculated first.

The path C to finish has a free float equal to its total float by definition. B to C has a free float of (14-11-1=) 2. Path A to B has a free float of (3-2-1=) 0.

The path I to finish has a free float equal to its total float by definition.

The path D to E has a free float equal to its total float by definition.

Question 22 - Develop the CPM chart (Hard)

Question:

Your project consists of nine tasks, A, B, C, D, E, F, G, H, and I. Task A is expected to take 20 days, task B is expected to take 4 days, task C is expected to take 12 days, task D is expected to take 23 days, task E is expected to take 4 days, task F is expected to take 6 days, task G is expected to take 18 days, task H is expected to take 6 days, and task I is expected to take 12 days. Task A, D, and G begin concurrently. Task B is preceded by task A. Task C is preceded by both task B and task H and ends the path. Task D precedes task E as does task H. Task F is preceded by task E and is the last task on the path. Task H is preceded by task G. Task I is preceded by task H and ends the path. Diagram the project using PDM, then prepare the CPM chart, identify the critical path, calculate all the dates, and calculate the floats.

Answer:

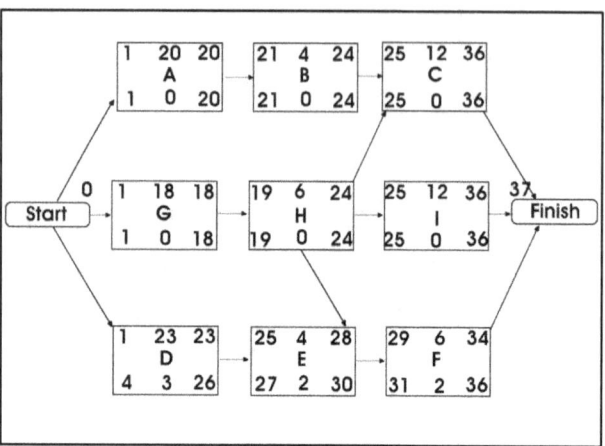

Process:

1. Design chart, placing letter or process name in the middle, and the duration on the top row in the middle.

2. There are five paths A-B-C, D-E-F, G-H-I, G-H-C, and G-H-E-F. A-B-C has a total duration of 20+4+12=36. G-H-I has a total duration of 18+6+12=36. D-E-F has a total duration of 23+4+6=33. G-H-C has a total duration of 18+6+12=36. G-H-E-F has a total duration of 18+6+4+6=34. A-B-C, G-H-I, and G-H-C are the

longest so they are the critical paths and to be calculated first.

3. Calculate the forward pass for the critical path A-B-C. This goes on the upper row of A, B, and C. Then calculate the forward pass for the critical path G-H-I. The result goes on the upper row of G, H, and I. Notice that C is common to A-B-C and G-H-C, and G-H is common to G-H-I and G-H-C. We have already calculated the tasks in G-H-C and therefore that path doesn't need to be calculated separately.

4. Perform the backward pass on the critical paths. To do this, copy the Early Start and Early Finish from the upper row into the Late Start and Late Finish. The Late Start and Late Finish go into the bottom row of the critical tasks.

5. Calculate the floats for the critical paths. This is always zero and is placed in the middle position of the bottom row of the critical tasks. The free float is also zero and therefore is ignored.

6. Calculate the forward paths for the non-critical tasks D, E, and F. This is placed on the top row of the non-critical tasks.

7. Calculate the backward paths for the non-critical tasks D, E, and F. This is placed on the bottom row of the non-critical tasks.

8. Calculate the total float for the non-critical tasks D, E, and F. This is placed in the middle position of the bottom row of the non-critical tasks.

9. The path F to finish has a free float equal to its total float by definition and doesn't have to be calculated again. E to F has a free float of (29-28-1=) 0. Path D to E has a free float of (25-23-1=) 1.

Question 23 - Calculate the values on the CPM chart (Hard)

Scenario:

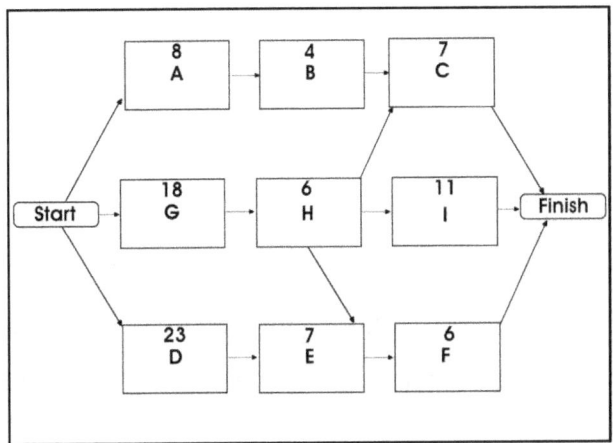

Question:

Given the above information, identify the critical path, calculate all the dates, and calculate the floats.

Answer:

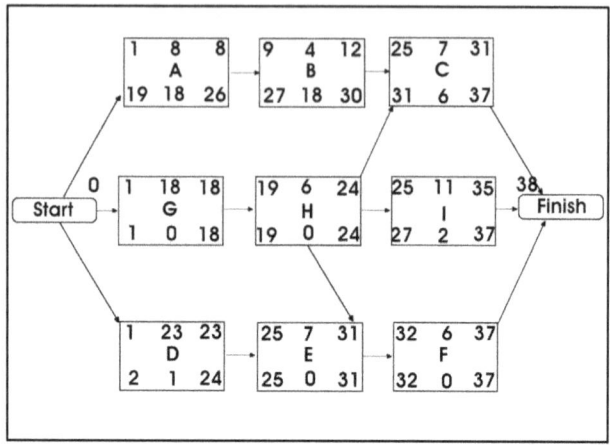

There are five paths A-B-C, D-E-F, G-H-I, G-H-C, and G-H-E-F. A-B-C has a total duration of 8+4+7=19. G-H-I has a total duration of 18+6+11=35. D-E-F has a total duration of 23+7+6=36. G-H-C has a total duration of 18+6+7=31. G-H-E-F has a total duration of 18+6+7+6=37. G-H-E-F is the longest so it is the critical path and to be calculated first.

The path C to finish has a free float equal to its total float by definition. B to C has a free float of (25-12-1=) 12. Path A to B has a free float of (9-8-1=) 0.

The path I to finish has a free float equal to its total float by definition.

The path D to E has a free float equal to its total float by definition.

Question 24 - Develop the CPM chart (Hard)

Question:

Your project consists of nine tasks, A, B, C, D, E, F, G, H, and I. Task A is expected to take 4 days, task B is expected to take 5 days, task C is expected to take 6 days, task D is expected to take 6 days, task E is expected to take 4 days, task F is expected to take 2 days, task G is expected to take 5 days, task H is expected to take 3 days, and task I is expected to take 3 days. Task A, D, and G begin concurrently. Task B is preceded by task A. Task C is preceded by both task B and task H and ends the path. Task D precedes task E as does task H. Task F is preceded by task E and is the last task on the path. Task H is preceded by task G. Task I is preceded by task H and ends the path. Diagram the project using PDM, then prepare the CPM chart, identify the critical path, calculate all the dates, and calculate the floats.

Answer:

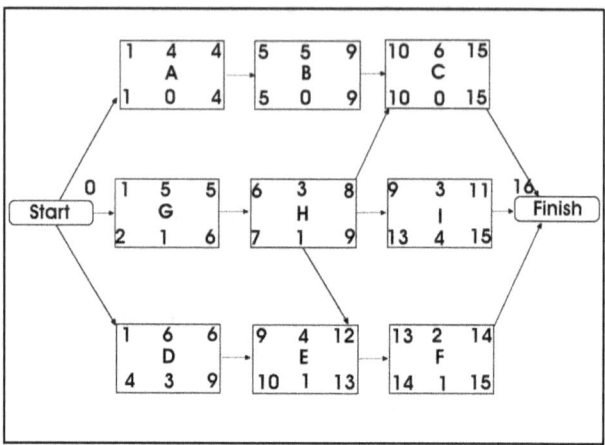

Process:

1. Design chart, placing letter or process name in the middle, and the duration on the top row in the middle.

2. There are five paths A-B-C, D-E-F, G-H-I, G-H-C, and G-H-E-F. A-B-C has a total duration of 4+5+6=15. G-H-I has a total duration of 5+3+3=11. D-E-F has a total duration of 6+4+2=12. G-H-C has a total duration of 5+3+6=14. G-H-E-F has a total duration of

5+3+4+2=14. A-B-C is the longest so it is the critical path and to be calculated first.

3. Calculate the forward pass for the critical path A-B-C. This goes on the upper row of tasks A, B, and C.

4. Perform the backward pass on the critical paths. To do this, copy the Early Start and Early Finish from the upper row into the Late Start and Late Finish. The Late Start and Late Finish go into the bottom row of the critical tasks.

5. Calculate the floats for the critical paths. This is always zero and is placed in the middle position of the bottom row of the critical tasks. The free float is also zero and therefore is ignored.

6. Calculate the forward paths for the non-critical tasks G, H, and I. This is placed on the top row of tasks G, H and I. Then calculate the forward paths for the non-critical tasks D, E, and F. This is placed on the top row of tasks D, E, and F.

7. Calculate the backward paths for the non-critical tasks G, H, and I. This is placed on the bottom row of tasks G, H, and I. Then calculate the backward paths for the non-critical tasks D, E, and F. This is placed on the bottom row of tasks D, E, and F.

8. Calculate the total float for tasks D, E, F, G, H, and I. This is placed in the middle position of the bottom row for each task.

9. The path I to finish has a free float equal to its total float by definition and doesn't have to be calculated again. H to I has a free float of (13-12-1=) 0. Path G to H has a free float of (6-5-1=) 0.

The path F to finish has a free float equal to its total float by definition and doesn't have to be calculated again. E to F has a free float of (13-12-1=) 0. Path D to E has a free float of (9-6-1=) 2.

Question 25 - Calculate the values on the CPM chart (Hard)

Scenario:

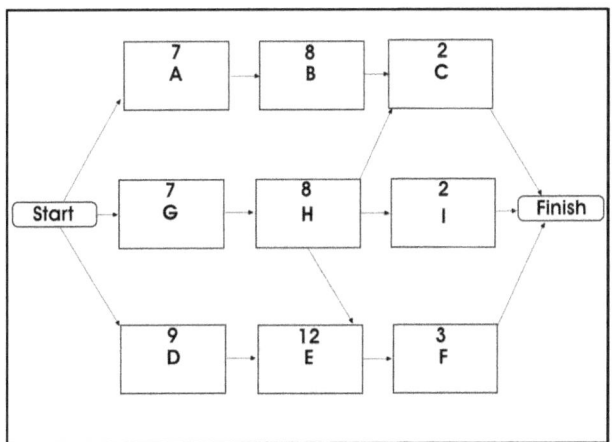

Question:

Given the above information, identify the critical path, calculate all the dates, and calculate the floats.

Answer:

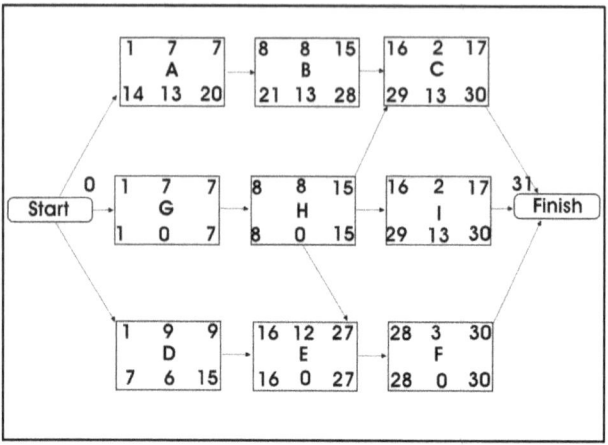

There are five paths A-B-C, D-E-F, G-H-I, G-H-C, and G-H-E-F. A-B-C has a total duration of 7+8+2=17. G-H-I has a total duration of 7+8+2=17. D-E-F has a total duration of 9+12+3=24. G-H-C has a total duration of 7+8+2=17. G-H-E-F has a total duration of 7+8+12+3=30. G-H-E-F is the longest so it is the critical path and to be calculated first.

The path C to finish has a free float equal to its total float by definition. B to C has a free float of (16-15-1=) 0. Path A to B has a free float of (8-7-1=) 0.

The path I to finish has a free float equal to its total float by definition.

The path D to E has a free float equal to its total float by definition.

Question:

You are running an agile project with three teams. Each sprint consists of 15 work days. Team 1 is responsible for sprints A, B, and C. Team 2 is responsible for sprints D, E, and F. And Team 3 is responsible for sprints G, H, and I. The sprints are labeled sequentially for each team (so the order for team 1 is sprint A, then sprint B, then sprint C) due to the nature of the work included in each sprint. Unfortunately, you discover that there is a relationship between the sprints across the teams. Sprint H must precede both sprint C and Sprint E (as well as sprint I).

Diagram the project using PDM, then prepare the CPM chart, identify the critical path, calculate all the dates, and calculate the floats.

Answer:

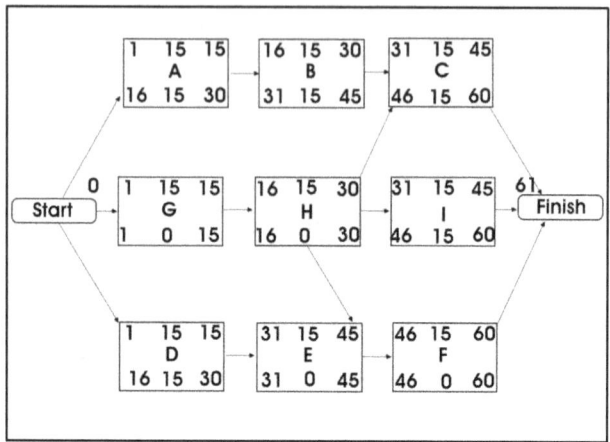

Process:

1. Design chart, placing letter or process name in the middle, and the duration on the top row in the middle.

2. There are five paths A-B-C, D-E-F, G-H-I, G-H-C, and G-H-E-F. A-B-C, D-E-F, G-H-I, and G-H-C have a total duration of 15+15+15=45. G-H-E-F has a total duration of 15+15+15+15=60. G-H-E-F is the longest so it is the critical paths and to be calculated first.

3. Calculate the forward pass for the critical path G-H-E-F. This goes on the upper row of tasks G, H, E, and F.

4. Perform the backward pass on the critical path. To do this, copy the Early Start and Early Finish from the upper row of the critical tasks into the Late Start and Late Finish. The Late Start and Late Finish go into the bottom row of the critical tasks.

5. Calculate the floats for the critical path. This is always zero and is placed in the middle position of the bottom row of the critical tasks. The free float is also zero and therefore is ignored.

6. Calculate the forward paths for the non-critical path A-B-C, and the non-critical tasks D and I. This is placed on the top row of the non-critical tasks.

7. Calculate the backward paths for the non-critical tasks D and I and the non-critical path A-B-C. This is placed on the bottom row of the tasks.

8. Calculate the total float for tasks A, B, C, D, and I. This is placed in the middle position of the bottom row for each task.

9. The task C and I to finish and task D to E have a free float equal to their total float by definition. There is no need to calculate the value again. B to C has a free float of (31-30-1=) 0. Path A to B has a free float of (16-15-1=) 0.

Question 27 - Calculate the values on the CPM chart (Hard)

Scenario:

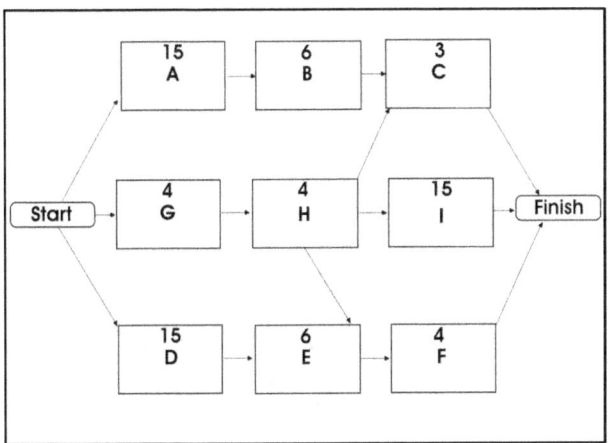

Question:

Given the above information, identify the critical path, calculate all the dates, and calculate the floats.

Answer:

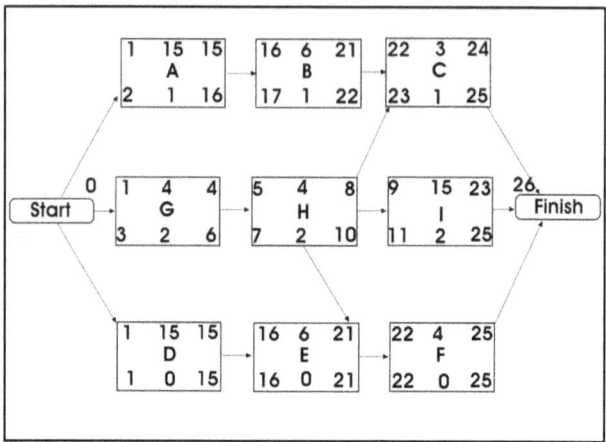

There are five paths A-B-C, D-E-F, G-H-I, G-H-C, and G-H-E-F. A-B-C has a total duration of 15+6+3=24. G-H-I has a total duration of 4+4+15=23. D-E-F has a total duration of 15+6+4=25. G-H-C has a total duration of 4+4+3=11. G-H-E-F has a total duration of 4+4+6+4=18. D-E-F is the longest so it is the critical path and to be calculated first.

The path C to finish has a free float equal to the total float by definition. B to C has a free float of (22-21-1=) 0. Path A to B has a free float of (16-15-1=) 0.

The path I to finish has a free float equal to its total float by definition. H to I has a free float of (9-8-1=) 0. Path G to H has a free float of (5-4-1=) 0.

Question 28 - Develop the CPM chart (Hard)

Question:

Your project consists of nine tasks, A, B, C, D, E, F, G, H, and I. Task A is expected to take 8 days, task B is expected to take 2 days, task C is expected to take 10 days, task D is expected to take 10 days, task E is expected to take 6 days, task F is expected to take 4 days, task G is expected to take 8 days, task H is expected to take 4 days, and task I is expected to take 8 days. Task A, D, and G begin concurrently. Task B is preceded by task A. Task C is preceded by both task B and task H and ends the path. Task D precedes task E as does task H. Task F is preceded by task E and is the last task on the path. Task H is preceded by task G. Task I is preceded by task H and ends the path. Diagram the project using PDM, then prepare the CPM chart, identify the critical path, calculate all the dates, and calculate the floats.

Answer:

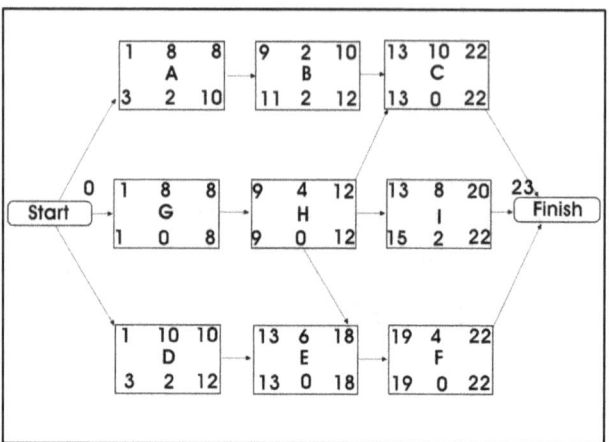

Process:

1. Design chart, placing letter or process name in the middle, and the duration on the top row in the middle.

2. There are five paths A-B-C, D-E-F, G-H-I, G-H-C, and G-H-E-F. A-B-C has a total duration of 8+2+10=20. G-H-I has a total duration of 8+4+8=20. D-E-F has a total duration of 10+6+4=20. G-H-C has a total duration of 8+4+10=22. G-H-E-F has a total duration of 8+4+6+4=22. G-H-C and G-H-E-F are the longest

so they are the critical paths and to be calculated first.

3. Calculate the forward pass for the critical path G-H-C. This goes on the upper row of tasks G, H, and C. Then calculate the forward pass for the critical path G-H-E-F. Notice that you only really need to calculate the tasks E and F because G and H are already completed. The result goes on the upper row of E and F.

4. Perform the backward pass on the critical paths. To do this, copy the Early Start and Early Finish from the upper row into the Late Start and Late Finish. The Late Start and Late Finish go into the bottom row of the critical tasks.

5. Calculate the floats for the critical paths. This is always zero and is placed in the middle position of the bottom row of the critical tasks. The free float is also zero and therefore is ignored.

6. Calculate the forward paths for the non-critical tasks A, B, D and I. This is placed on the top row of the tasks.

7. Calculate the backward paths for the non-critical tasks A, B, D and I. This is placed on the bottom row of the tasks.

8. Calculate the total float for tasks A, B, D and I. This is placed in the middle position of the bottom row of the tasks.

9. The path B to C has a free float equal to its total float by definition and doesn't have to be calculated again. A to B has a free float of (9-8-1=) 0.

The path I to finish has a free float equal to its total float by definition and doesn't have to be calculated again.

The path D to E has a free float equal to its total float by definition and doesn't have to be calculated again.

Question 29 - Calculate the values on the CPM chart (Hard)

Scenario:

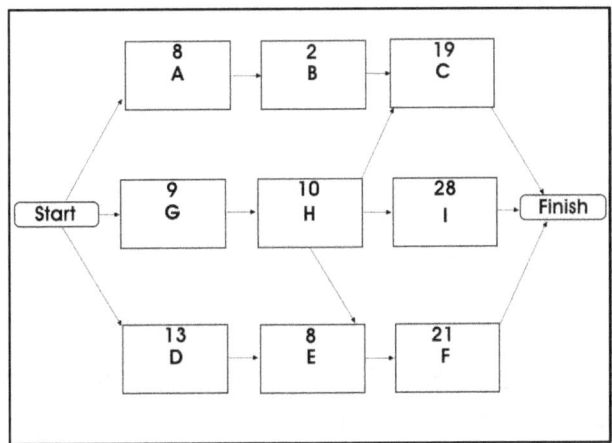

Question:

Given the above information, identify the critical path, calculate all the dates, and calculate the floats.

Answer:

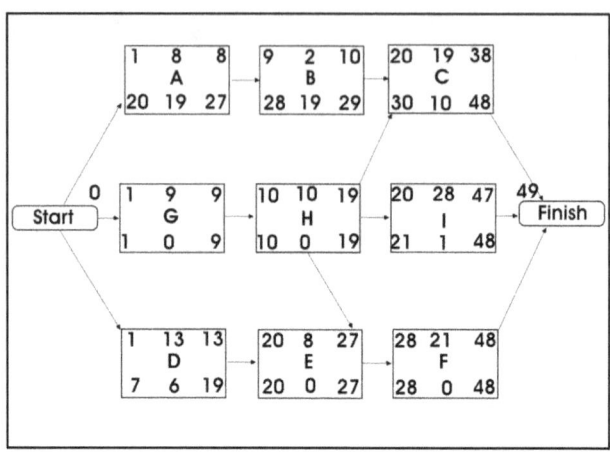

There are five paths A-B-C, D-E-F, G-H-I, G-H-C, and G-H-E-F. A-B-C has a total duration of 8+2+19=29. G-H-I has a total duration of 9+10+28=47. D-E-F has a total duration of 13+8+21=42. G-H-C has a total duration of 9+10+19=38. G-H-E-F has a total duration of 9+10+8+21=48. G-H-E-F is the longest so it is the critical path and to be calculated first.

The path C to finish has a free float equal to its total float by definition. B to C has a free float of (20-10-1=) 9. Path A to B has a free float of (9-8-1=) 0.

The path I to finish has a free float equal to its total float by definition.

The path D to E has a free float equal to its total float by definition.

Question 30 - Develop the CPM chart (Hard)

Question:

Your project consists of nine tasks, A, B, C, D, E, F, G, H, and I. Task A is expected to take 10 days, task B is expected to take 19 days, task C is expected to take 4 days, task D is expected to take 13 days, task E is expected to take 16 days, task F is expected to take 5 days, task G is expected to take 7 days, task H is expected to take 23 days, and task I is expected to take 25 days. Task A, D, and G begin concurrently. Task B is preceded by task A. Task C is preceded by both task B and task H and ends the path. Task D precedes task E as does task H. Task F is preceded by task E and is the last task on the path. Task H is preceded by task G. Task I is preceded by task H and ends the path. Diagram the project using PDM, then prepare the CPM chart, identify the critical path, calculate all the dates, and calculate the floats.

Answer:

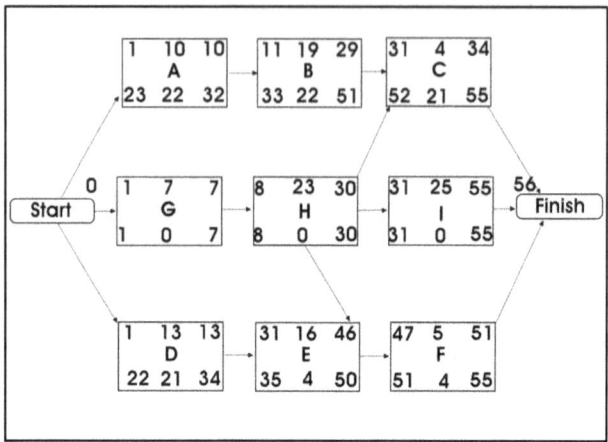

Process:

1. Design chart, placing letter or process name in the middle, and the duration on the top row in the middle.

2. There are five paths A-B-C, D-E-F, G-H-I, G-H-C, and G-H-E-F. A-B-C has a total duration of 10+19+4=33. G-H-I has a total duration of 7+23+25=55. D-E-F has a total duration of 13+16+5=34. G-H-C has a total duration of 7+23+4=34. G-H-E-F has a total duration of

7+23+16+5=51. G-H-I is the longest so it is the critical path and to be calculated first.

3. Calculate the forward pass for the critical path G-H-I. This goes on the upper row of tasks G, H, and I.

4. Perform the backward pass on the critical path. To do this, copy the Early Start and Early Finish from the upper row of the critical tasks into the Late Start and Late Finish. The Late Start and Late Finish go into the bottom row of the critical tasks.

5. Calculate the floats for the critical path. This is always zero and is placed in the middle position of the bottom row of the critical tasks. The free float is also zero and therefore is ignored.

6. Calculate the forward paths for the non-critical tasks A, B and C. This is placed on the top row of A, B, and C.

Calculate the forward paths for the non-critical tasks D, E and F. This is placed on the top row of D, E, and F.

7. Calculate the backward paths for the non-critical tasks A, B and C. This is placed on the bottom row of tasks A, B, and C.

Calculate the backward paths for the non-critical tasks D, E, and F. This is placed on the bottom row of tasks D, E, and F.

8. Calculate the total float for tasks A, B, C, D, E, and F. This is placed in the middle position of the bottom row of tasks A, B, C, D, E, and F.

9. The path C to finish has a free float equal to its total float by definition and doesn't have to be calculated again. B to C has a free float of (31-29-1=) 1. Path A to B has a free float of (11-10-1=) 0.

The path F to finish has a free float equal to its total float by definition and doesn't have to be calculated again. E to F has a free float of (47-46-1=) 0. Path D to E has a free float of (31-13-1=) 17.

Question 31 - Calculate the values on the CPM chart (Very Hard)

Scenario:

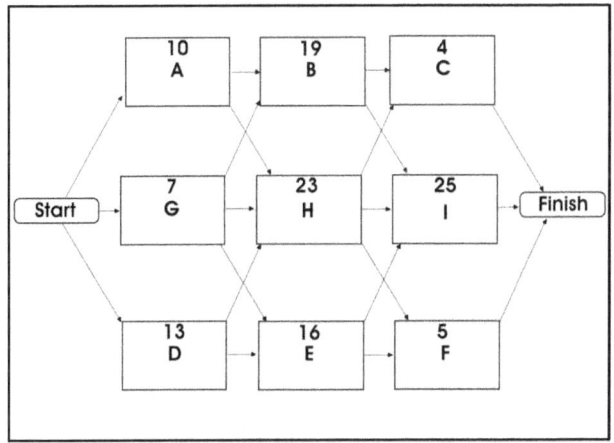

Question:

Given the above information, identify the critical path, calculate all the dates, and calculate the floats.

Answer:

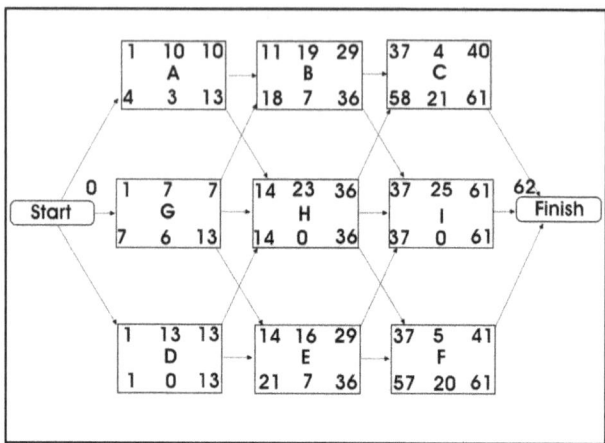

There are seventeen paths A-B-C, D-E-F, G-H-I, A-B-I, A-H-C, A-H-I, A-H-F, G-B-C, G-B-I, G-H-C, G-H-F, G-E-I, G-E-F, D-H-C, D-H-I, D-H-F, and D-E-I. A-B-C has a total duration of 10+19+4=33. G-H-I has a total duration of 7+23+25=55. D-E-F has a total duration of 13+16+5=34. A-B-I has a total duration of 10+19+25=54. A-H-C has a total duration of 10+23+4=37. A-H-I has a total duration of 10+23+25=58. A-H-F has a total duration of 10+23+5=38. G-B-C has a total duration of 7+19+4=30. G-B-I has a total duration of 7+19+25=51. G-H-C has a total duration of 7+23+4=34. G-H-F has a total duration of 7+23+5=35. G-E-I has a total duration of

7+16+25=48. G-E-F has a total duration of 7+16+5=28. D-H-C has a total duration of 13+23+4=40. D-H-I has a total duration of 13+23+25=61. D-H-F has a total duration of 13+23+5=41. D-E-I has a total duration of 13+16+25=54. D-H-I is the longest so it is the critical path and to be calculated first.

The path C to finish has a free float equal to its total float by definition. B to C has a free float of (37-29-1=) 7. B to I has a free float of (37-29-1=) 7. (Task B is generally held to have a free float of the minimum of these two). Path A to B has a free float of (11-10-1=) 0. Path A to H has a free float of (14-10-1=) 3. (Task A is generally held to have a free float of the minimum of these two).

The path G to H has a free float equal to its total float by definition. Path G to B has a free float of (11-7-1=) 3. Path G to E has a free float of (14-7-1=) 6. (Task G is generally held to have a free float of the minimum of these three).

The path F to finish has a free float equal to its total float by definition. The path E to I has a free float equal to its total float by definition. Path E to F has a free float of (37-29-1=) 7. (Task E is generally held to have a free float of the minimum of these two).

Question:

Your project consists of nine tasks, A, B, C, D, E, F, G, H, and I. Task A is expected to take 25 days, task B is expected to take 22 days, task C is expected to take 8 days, task D is expected to take 17 days, task E is expected to take 12 days, task F is expected to take 3 days, task G is expected to take 12 days, task H is expected to take 7 days, and task I is expected to take 9 days. Task A, D, and G begin concurrently. Task B is preceded by task A and G. Task C is preceded by task B and task H and also ends the path. Task D precedes task E and task H. Task F is preceded by task E and H and is the last task on the path. Task H is preceded by task G, task A, and task D. Task I is preceded by task H, task B, and task E. It also ends the path. Diagram the project using PDM, then prepare the CPM chart, identify the critical path, calculate all the dates, and calculate the floats.

Answer:

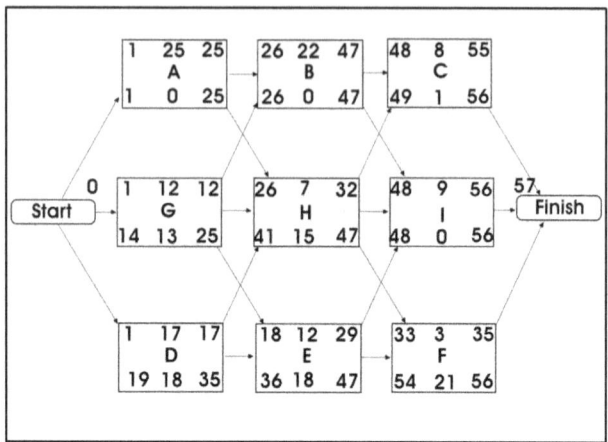

Process:

1. Design chart, placing letter or process name in the middle, and the duration on the top row in the middle.

2. There are seventeen paths A-B-C, D-E-F, G-H-I, A-B-I, A-H-C, A-H-I, A-H-F, G-B-C, G-B-I, G-H-C, G-H-F, G-E-I, G-E-F, D-H-C, D-H-I, D-H-F, and D-E-I. A-B-C has a total duration of 25+22+8=55. G-H-I has a total duration of 12+7+9=28. D-E-F has a total duration of 17+12+3=32. A-B-I has a total duration of 25+22+9=56. A-H-C has a total

duration of 25+7+8=40. A-H-I has a total duration of 25+7+9=41. A-H-F has a total duration of 25+7+3=35. G-B-C has a total duration of 12+22+8=42. G-B-I has a total duration of 12+22+9=43. G-H-C has a total duration of 12+7+8=27. G-H-F has a total duration of 12+7+3=22. G-E-I has a total duration of 12+12+9=33. G-E-F has a total duration of 12+12+3=27. D-H-C has a total duration of 17+7+8=32. D-H-I has a total duration of 17+7+9=33. D-H-F has a total duration of 17+7+3=27. D-E-I has a total duration of 17+12+9=38. A-B-I is the longest so it is the critical path and to be calculated first.

3. Calculate the forward pass for the critical path A-B-I. This goes on the upper row of tasks A, B, and I.

4. Perform the backward pass on the critical path. To do this, copy the Early Start and Early Finish from the upper row of the critical tasks into the Late Start and Late Finish. The Late Start and Late Finish go into the bottom row of the critical tasks.

5. Calculate the floats for the critical path. This is always zero and is placed in the middle position of the bottom row of the critical tasks. The free float is also zero and therefore is ignored.

6. Calculate the forward paths for the non-critical task C. This is placed on the top row of C.

Calculate the forward paths for the non-critical tasks D, E and F. This is placed on the top row of D, E, and F.

Calculate the forward paths for the non-critical tasks G, and H. This is placed on the top row of G, and H.

7. Calculate the backward paths for the non-critical task C. This is placed on the bottom row of task C.

Calculate the backward paths for the non-critical tasks D, E, and F. This is placed on the bottom row of tasks D, E, and F.

Calculate the backward paths for the non-critical tasks G, and H. This is placed on the bottom row of tasks G, and H.

8. Calculate the total float for tasks C, D, E, F, G, and H. This is placed in the middle position of the bottom row of tasks C, D, E, F, G, and H.

9. The path C to finish has a free float equal to its total float by definition and doesn't have to be calculated again.

The path F to finish has a free float equal to its total float by definition and doesn't have to be calculated again. E to F has a free float of (33-29-1=) 3. Path D to E has a free float of (18-17-1=) 0.

The path H to I has a free float equal to its total float by definition and doesn't have to be calculated again. G to H has a free float of (26-12-1=) 13.

Question 33 - Calculate the values on the CPM chart (Very Hard)

Scenario:

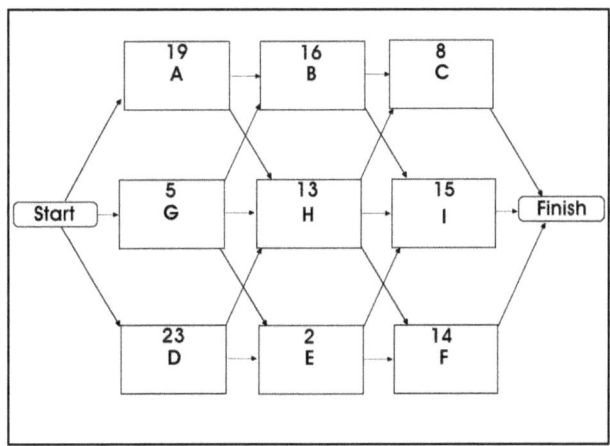

Question:

Given the above information, identify the critical path, calculate all the dates, and calculate the floats.

Answer:

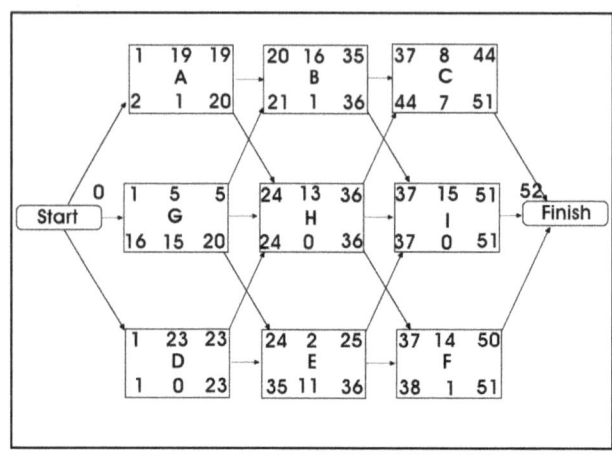

There are seventeen paths A-B-C, D-E-F, G-H-I, A-B-I, A-H-C, A-H-I, A-H-F, G-B-C, G-B-I, G-H-C, G-H-F, G-E-I, G-E-F, D-H-C, D-H-I, D-H-F, and D-E-I. A-B-C has a total duration of 19+16+8=43. G-H-I has a total duration of 5+13+15=33. D-E-F has a total duration of 23+2+14=39. A-B-I has a total duration of 19+16+15=50. A-H-C has a total duration of 19+13+8=40. A-H-I has a total duration of 19+13+15=47. A-H-F has a total duration of 19+13+14=46. G-B-C has a total duration of 5+16+8=29. G-B-I has a total duration of 5+16+15=36. G-H-C has a total duration of 5+13+8=26. G-H-F has a total duration of 5+13+14=32. G-E-I has a total duration of

5+2+15=22. G-E-F has a total duration of 5+2+14=21. D-H-C has a total duration of 23+13+8=44. D-H-I has a total duration of 23+13+15=51. D-H-F has a total duration of 23+13+14=50. D-E-I has a total duration of 23+2+15=40. D-H-I is the longest so it is the critical path and to be calculated first.

The path C to finish has a free float equal to its total float by definition. B to C has a free float of (37-35-1=) 1. B to I has a free float of (37-35-1=) 1. (Task B is generally held to have a free float of the minimum of these two). Path A to B has a free float of (20-19-1=) 0. Path A to H has a free float of (24-19-1=) 4. (Task A is generally held to have a free float of the minimum of these two).

The path G to H has a free float of (24-5-1=) 18. Path G to B has a free float of (20-5-1=) 15. Path G to E has a free float of (24-5-1=) 18. (Task G is generally held to have a free float of the minimum of these three).

The path F to finish has a free float equal to its total float by definition. The path E to I has a free float of (37-25-1=) 11. Path E to F has a free float of (37-25-1=) 11. (Task E is generally held to have a free float of the minimum of these two).

Question 34 - Develop the CPM chart (Very Hard)

Question:

Your project consists of nine tasks, A, B, C, D, E, F, G, H, and I. Task A is expected to take 7 days, task B is expected to take 6 days, task C is expected to take 5 days, task D is expected to take 19 days, task E is expected to take 12 days, task F is expected to take 13 days, task G is expected to take 13 days, task H is expected to take 4 days, and task I is expected to take 13 days. Task A, D, and G begin concurrently. Task B is preceded by task A and G. Task C is preceded by task B and task H and also ends the path. Task D precedes task E and task H. Task F is preceded by task E and H and is the last task on the path. Task H is preceded by task G, task A, and task D. Task I is preceded by task H, task B, and task E. It also ends the path. Diagram the project using PDM, then prepare the CPM chart, identify the critical path, calculate all the dates, and calculate the floats.

Answer:

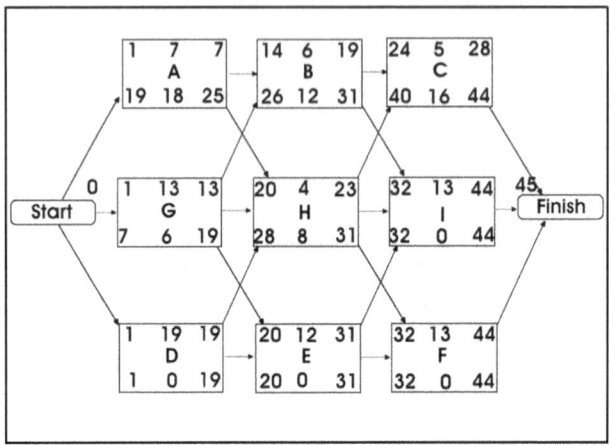

Process:

1. Design chart, placing letter or process name in the middle, and the duration on the top row in the middle.

2. There are seventeen paths A-B-C, D-E-F, G-H-I, A-B-I, A-H-C, A-H-I, A-H-F, G-B-C, G-B-I, G-H-C, G-H-F, G-E-I, G-E-F, D-H-C, D-H-I, D-H-F, and D-E-I. A-B-C has a total duration of 7+6+5=18. G-H-I has a total duration of 13+4+13=30. D-E-F has a total duration of 19+12+13=44. A-B-I has a total duration of 7+6+13=26. A-H-C has a total duration

of 7+4+5=16. A-H-I has a total duration of 7+4+13=24. A-H-F has a total duration of 7+4+13=24. G-B-C has a total duration of 13+6+5=24. G-B-I has a total duration of 13+6+13=32. G-H-C has a total duration of 13+4+5=22. G-H-F has a total duration of 13+4+13=30. G-E-I has a total duration of 13+12+13=38. G-E-F has a total duration of 13+12+13=38. D-H-C has a total duration of 19+4+5=28. D-H-I has a total duration of 19+4+13=36. D-H-F has a total duration of 19+4+13=36. D-E-I has a total duration of 19+12+13=44. D-E-F and D-E-I are the longest so they are the critical paths and to be calculated first.

3. Calculate the forward pass for the critical path D-E-F. This goes on the upper row of tasks D, E, and F. Calculate the forward pass for task I (i.e. path D-E-I). This goes on the upper row of I.

4. Perform the backward pass on the critical paths. To do this, copy the Early Start and Early Finish from the upper row into the Late Start and Late Finish. The Late Start and Late Finish go into the bottom row of tasks D, E, F, and I.

5. Calculate the floats for the critical paths. This is always zero and is placed in the middle position of the bottom row of tasks D, E, F, and I. The free float is also zero and therefore is ignored.

6. Calculate the forward paths for the non-critical tasks A, B and C. This is placed on the top row of tasks A, B, and C.

Calculate the forward paths for the non-critical tasks G, and H. This is placed on the top row of tasks G, and H.

7. Calculate the backward paths for the non-critical tasks A, B and C. This is placed on the bottom row of tasks A, B, and C.

Calculate the backward paths for the non-critical tasks G, and H. This is placed on the bottom row of tasks G, and H.

8. Calculate the total float for tasks A, B, C, G, and H. This is placed in the middle position of the bottom row of tasks A, B, C, G, and H.

9. The path C to finish has a free float equal to its total float by definition and doesn't have to be calculated again. B to C has a free float of (24-19-1=) 4. B to I has a free float of (32-19-1=) 12. (Task B is generally held to have a free float of the minimum of these two). Path A to B has a free float of (14-7-1=) 6. A to H has a free float of (20-7-1=) 12. (Task A is generally held to have a free float of the minimum of these two).

The path H to C has a free float of (24-23-1=) 0. Path H to I has a free float of (32-23-1=) 8. Path H to F has a free float of (32-23-1=) 8. (Task H is generally held to have a free float of the minimum of these three). The path G to H has a free float of (20-13-1=) 6. Path G to B has a free float of (14-13-1=) 0. Path G to E has a free float of (20-13-1=) 6. (Task G is generally held to have a free float of the minimum of these three).

Question 35 - Calculate the values on the CPM chart (Very Hard)

Scenario:

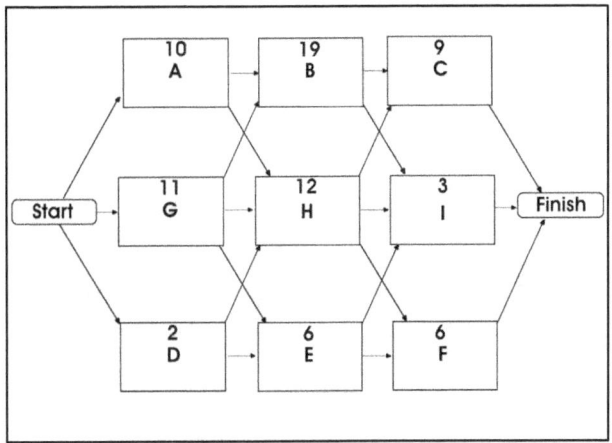

Question:

Given the above information, identify the critical path, calculate all the dates, and calculate the floats.

Answer:

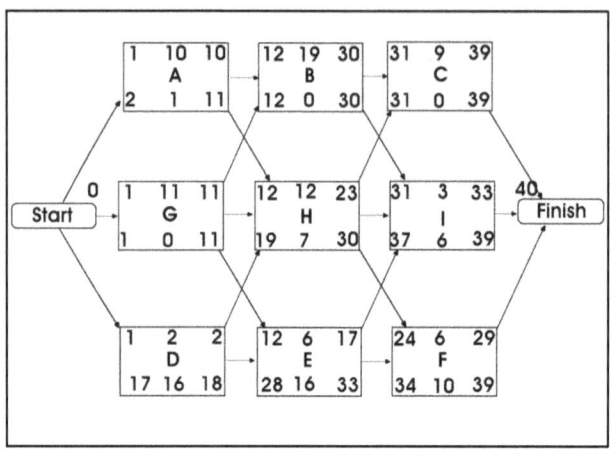

There are seventeen paths A-B-C, D-E-F, G-H-I, A-B-I, A-H-C, A-H-I, A-H-F, G-B-C, G-B-I, G-H-C, G-H-F, G-E-I, G-E-F, D-H-C, D-H-I, D-H-F, and D-E-I. A-B-C has a total duration of 10+19+9=38. G-H-I has a total duration of 11+12+3=26. D-E-F has a total duration of 2+6+6=14. A-B-I has a total duration of 10+19+3=32. A-H-C has a total duration of 10+12+9=31. A-H-I has a total duration of 10+12+3=25. A-H-F has a total duration of 10+12+6=28. G-B-C has a total duration of 11+19+9=39. G-B-I has a total duration of 11+19+3=33. G-H-C has a total duration of 11+12+9=32. G-H-F has a total duration of 11+12+6=29. G-E-I has a total duration of

11+6+3=20. G-E-F has a total duration of 11+6+6=23. D-H-C has a total duration of 2+12+9=23. D-H-I has a total duration of 2+12+3=17. D-H-F has a total duration of 2+12+6=20. D-E-I has a total duration of 2+6+3=11. G-B-C is the longest so it is the critical path and to be calculated first.

The path A to B has a free float of (12-10-1=) 1. Path A to H has a free float of (12-10-1=) 1. (Task A is generally held to have a free float of the minimum of these two).

The path I to finish has a free float equal to its total float by definition and doesn't have to be calculated again. The path H to C has a free float of (31-23-1=) 7. Path H to I has a free float of (31-23-1=) 7. Path H to F has a free float of (24-23-1=) 0. (Task H is generally held to have a free float of the minimum of these three).

The path F to finish has a free float equal to its total float by definition. The path E to I has a free float of (31-17-1=) 13. Path E to F has a free float of (24-17-1=) 6. (Task E is generally held to have a free float of the minimum of these three).

The path D to H has a free float of (12-2-1=) 9. Path D to E has a free float of (12-2-1=) 9. (Task D therefore has a free float of 9).

Question 36 - Develop the CPM chart (Very Hard)

Question:

Your project consists of nine tasks, A, B, C, D, E, F, G, H, and I. Task A is expected to take 7 days, task B is expected to take 25 days, task C is expected to take 18 days, task D is expected to take 19 days, task E is expected to take 10 days, task F is expected to take 23 days, task G is expected to take 11 days, task H is expected to take 23 days, and task I is expected to take 14 days. Task A, D, and G begin concurrently. Task B is preceded by task A and G. Task C is preceded by task B and task H and also ends the path. Task D precedes task E and task H. Task F is preceded by task E and H and is the last task on the path. Task H is preceded by task G, task A, and task D. Task I is preceded by task H, task B, and task E. It also ends the path. Diagram the project using PDM, then prepare the CPM chart, identify the critical path, calculate all the dates, and calculate the floats.

Answer:

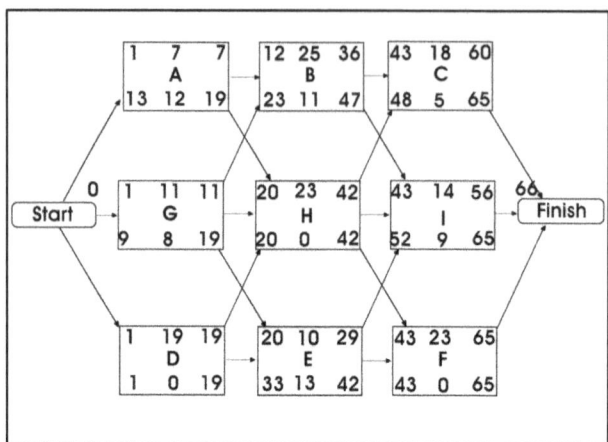

Process:

1. Design chart, placing letter or process name in the middle, and the duration on the top row in the middle.

2. There are seventeen paths A-B-C, D-E-F, G-H-I, A-B-I, A-H-C, A-H-I, A-H-F, G-B-C, G-B-I, G-H-C, G-H-F, G-E-I, G-E-F, D-H-C, D-H-I, D-H-F, and D-E-I. A-B-C has a total duration of 7+25+18=50. G-H-I has a total duration of 11+23+14=48. D-E-F has a total duration of 19+10+23=52. A-B-I has a total duration of 7+25+14=46. A-H-C has a total duration of 7+23+18=48. A-H-I has a total duration of 7+23+14=44. A-H-F has a total duration of

172

7+23+23=53. G-B-C has a total duration of 11+25+18=54. G-B-I has a total duration of 11+25+14=50. G-H-C has a total duration of 11+23+18=52. G-H-F has a total duration of 11+23+23=57. G-E-I has a total duration of 11+10+14=35. G-E-F has a total duration of 11+10+23=44. D-H-C has a total duration of 19+23+18=60. D-H-I has a total duration of 19+23+14=56. D-H-F has a total duration of 19+23+23=65. D-E-I has a total duration of 19+10+14=43. D-H-F is the longest so it is the critical path and to be calculated first.

3. Calculate the forward pass for the critical path D-H-F. This goes on the upper row of tasks D, H, and F.

4. Perform the backward pass on the critical path. To do this, copy the Early Start and Early Finish from the upper row into the Late Start and Late Finish. The Late Start and Late Finish go into the bottom row of tasks D, H, and F.

5. Calculate the floats for the critical path. This is always zero and is placed in the middle

position of the bottom row of tasks D, H, and F. The free float is also zero and therefore is ignored.

6. Calculate the forward paths for the non-critical tasks A, B and C. This is placed on the top row of A, B, and C.

Calculate the forward paths for the non-critical tasks G, and I. This is placed on the top row of G, and I.

Calculate the forward paths for the non-critical task E. This is placed on the top row of E.

7. Calculate the backward paths for the non-critical tasks A, B and C. This is placed on the bottom row of tasks A, B, and C.

Calculate the backward paths for the non-critical tasks G, and I. This is placed on the bottom row of tasks G, and I.

Calculate the backward paths for the non-critical task E. This is placed on the bottom row of E.

8. Calculate the total float for tasks A, B, C, E, G, and I. This is placed in the middle position of the bottom row of tasks A, B, C, E, G, and I.

9. The path C to finish has a free float equal to its total float by definition and doesn't have to be calculated again. The path B to C has a free float of (43-36-1=) 6. Path B to I has a free float of (43-36-1=) 6. (Task B is generally held to have a free float of the minimum of these two). Path A to B has a free float of (12-7-1=) 4. Path A to H has a free float of (20-7-1=) 12. (Task A is generally held to have a free float of the minimum of these two).

The path I to finish has a free float equal to its total float by definition and doesn't have to be calculated again. Path G to B has a free float of (12-11-1=) 0. Path G to H has a free float of (20-11-1=) 8. . Path G to E has a free float of (20-11-1=) 8. (Task G is generally held to have a free float of the minimum of these three).

The path E to F has a free float of (43-29-1=) 13. Path E to I has a free float of (43-29-1=) 13. (Task E is generally held to have a free float of the minimum of these two).

Question 37 - Calculate the values on the CPM chart (Very Hard)

Scenario:

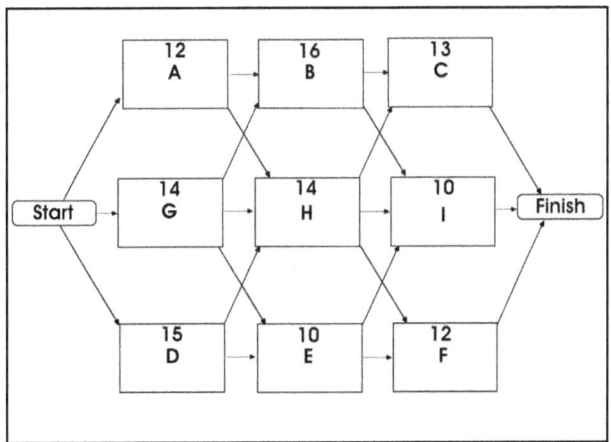

Question:

Given the above information, identify the critical path, calculate all the dates, and calculate the floats.

Answer:

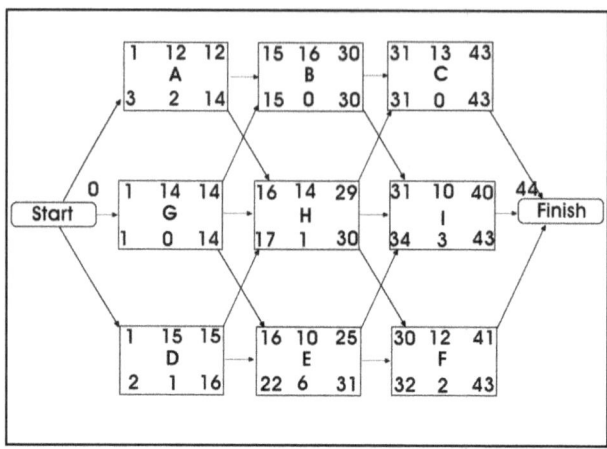

There are seventeen paths A-B-C, D-E-F, G-H-I, A-B-I, A-H-C, A-H-I, A-H-F, G-B-C, G-B-I, G-H-C, G-H-F, G-E-I, G-E-F, D-H-C, D-H-I, D-H-F, and D-E-I. A-B-C has a total duration of 12+16+13=41. G-H-I has a total duration of 14+14+10=38. D-E-F has a total duration of 15+10+12=37. A-B-I has a total duration of 12+16+10=38. A-H-C has a total duration of 12+14+13=39. A-H-I has a total duration of 12+14+10=36. A-H-F has a total duration of 12+14+12=38. G-B-C has a total duration of 14+16+13=43. G-B-I has a total duration of 14+16+10=40. G-H-C has a total duration of 14+14+13=34. G-H-F has a total duration of 14+14+12=40. G-E-I has a total duration of 14+10+10=34. G-E-F has a total duration of 14+10+12=36. D-H-C has a total duration of 15+14+13=42. D-H-I has a total duration of 15+14+10=39. D-H-F has a total duration of 15+14+12=41. D-E-I has a total duration of 15+10+10=35. G-B-C is the longest so it is the critical path and to be calculated first.

The path A to B has a free float of (15-12-1=) 2. Path A to H has a free float of (16-12-1=) 3. (Task A is generally held to have a free float of the minimum of these two).

The path I to finish has a free float equal to its total float by definition. Path H to C has a free float of (31-29-1=) 1. Path H to I has a free float of (31-29-1=) 1. Path H to F has a free float of (30-29-1=) 0. (Task H is generally held to have a free float of the minimum of these three).

The path F to finish has a free float equal to its total float by definition. The path E to I has a free float of (31-25-1=) 5. Path E to F has a free float of (30-25-1=) 4. (Task E is generally held to have a free float of the minimum of these two). The path D to H has a free float of (16-15-1=) 0. Path D to E has a free float of (16-15-1=) 0. (Task D is generally held to have a free float of the minimum of these two).

Question 38 - Develop the CPM chart (Very Hard)

Question:

Your project consists of nine tasks, A, B, C, D, E, F, G, H, and I. Task A is expected to take 12 days, task B is expected to take 16 days, task C is expected to take 13 days, task D is expected to take 15 days, task E is expected to take 10 days, task F is expected to take 12 days, task G is expected to take 12 days, task H is expected to take 14 days, and task I is expected to take 10 days. Task A, D, and G begin concurrently. Task B is preceded by task A and G. Task C is preceded by task B and task H and also ends the path. Task D precedes task E and task H. Task F is preceded by task E and H and is the last task on the path. Task H is preceded by task G, task A, and task D. Task I is preceded by task H, task B, and task E. It also ends the path. Diagram the project using PDM, then prepare the CPM chart, identify the critical path, calculate all the dates, and calculate the floats.

Answer:

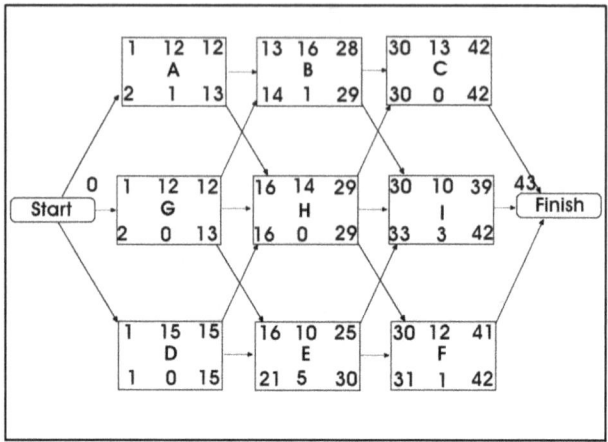

Process:

1. Design chart, placing letter or process name in the middle, and the duration on the top row in the middle.

2. There are seventeen paths A-B-C, D-E-F, G-H-I, A-B-I, A-H-C, A-H-I, A-H-F, G-B-C, G-B-I, G-H-C, G-H-F, G-E-I, G-E-F, D-H-C, D-H-I, D-H-F, and D-E-I. A-B-C has a total duration of 12+16+13=41. G-H-I has a total duration of 12+14+10=36. D-E-F has a total duration of 15+10+12=37. A-B-I has a total duration of 12+16+10=38. A-H-C has a total

duration of 12+14+13=39. A-H-I has a total duration of 12+14+10=36. A-H-F has a total duration of 12+14+12=38. G-B-C has a total duration of 12+16+13=41. G-B-I has a total duration of 12+16+10=38. G-H-C has a total duration of 12+14+13=39. G-H-F has a total duration of 12+14+12=38. G-E-I has a total duration of 12+10+10=32. G-E-F has a total duration of 12+10+12=34. D-H-C has a total duration of 15+14+13=42. D-H-I has a total duration of 15+14+10=39. D-H-F has a total duration of 15+14+12=41. D-E-I has a total duration of 15+10+10=35. D-H-C is the longest so it is the critical path and to be calculated first.

3. Calculate the forward pass for the critical path D-H-C. This goes on the upper row of tasks D, H, and C.

4. Perform the backward pass on the critical path. To do this, copy the Early Start and Early Finish from the upper row of tasks D, H, and C into the Late Start and Late Finish. The Late Start and Late Finish go into the bottom row of tasks D, H, and C.

5. Calculate the floats for the critical path. This is always zero and is placed in the middle position of the bottom row of tasks D, H, and C. The free float is also zero and therefore is ignored.

6. Calculate the forward paths for the non-critical tasks A, and B. This is placed on the top row of A, and B.

Calculate the forward paths for the non-critical tasks E and F. This is placed on the top row of E, and F.

Calculate the forward paths for the non-critical tasks G and I. This is placed on the top row of G, and I.

7. Calculate the backward paths for the non-critical tasks A, and B. This is placed on the bottom row of tasks A, and B.

Calculate the backward paths for the non-critical tasks E, and F. This is placed on the bottom row of tasks E, and F.

Calculate the backward paths for the non-critical tasks G, and I. This is placed on the bottom row of tasks G, and I.

8. Calculate the total float for tasks A, B, E, F, G, and I. This is placed in the middle position of the bottom row of tasks A, B, E, F, G, and I.

9. The path B to C has a free float of (30-28-1=) 1. The path B to I has a free float of (30-28-1=) 1. (Task B is generally held to have a free float of the minimum of these two). Path A to B has a free float of (13-12-1=) 0. Path A to H has a free float of (16-12-1=) 3. (Task A is generally held to have a free float of the minimum of these two).

The path I to finish has a free float equal to its total float by definition and doesn't have to be calculated again. G to B has a free float of (13-12-1=) 0. G to H has a free float of (16-12-1=) 3. G to E has a free float of (16-12-1=) 3. (Task G is generally held to have a free float of the minimum of these three).

The path F to finish has a free float equal to its total float by definition and doesn't have to be calculated again. E to I has a free float of (30-25-1=) 4. Path E to F has a free float of (30-25-1=) 4. (Task E is generally held to have a free float of the minimum of these two).

Question 39 - Calculate the values on the CPM chart (Very Hard)

Scenario:

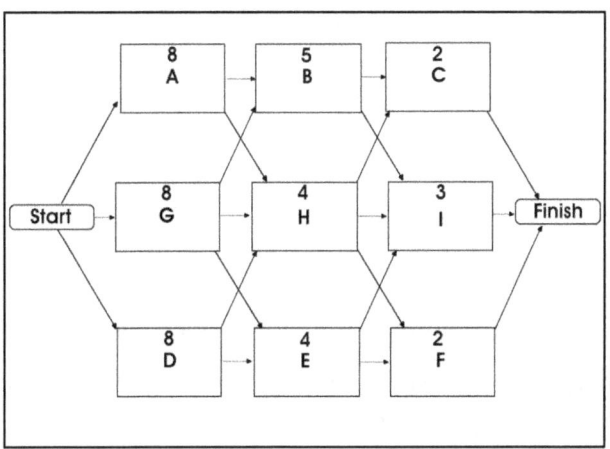

Question:

Given the above information, identify the critical path, calculate all the dates, and calculate the floats.

Answer:

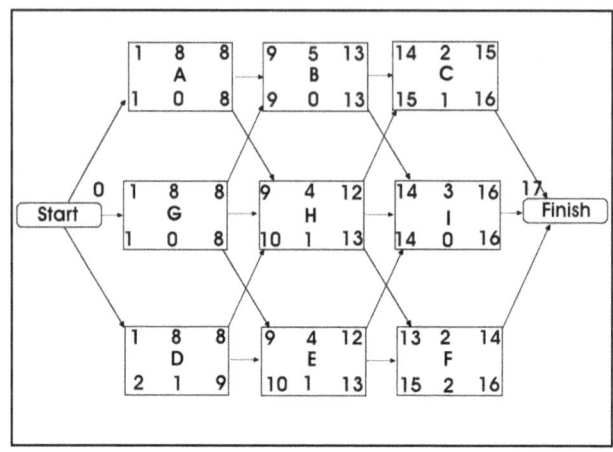

There are seventeen paths A-B-C, D-E-F, G-H-I, A-B-I, A-H-C, A-H-I, A-H-F, G-B-C, G-B-I, G-H-C, G-H-F, G-E-I, G-E-F, D-H-C, D-H-I, D-H-F, and D-E-I. A-B-C has a total duration of 8+5+2=15. G-H-I has a total duration of 8+4+3=15. D-E-F has a total duration of 8+4+2=14. A-B-I has a total duration of 8+5+3=16. A-H-C has a total duration of 8+4+2=14. A-H-I has a total duration of 8+4+3=15. A-H-F has a total duration of 8+4+2=14. G-B-C has a total duration of 8+5+2=15. G-B-I has a total duration of 8+5+3=16. G-H-C has a total duration of 8+4+2=14. G-H-F has a total duration of 8+4+2=14. G-E-I has a total duration of

8+4+3=15. G-E-F has a total duration of 8+4+2=14. D-H-C has a total duration of 8+4+2=14. D-H-I has a total duration of 8+4+3=15. D-H-F has a total duration of 8+4+2=14. D-E-I has a total duration of 8+4+3=15. A-B-I and G-B-I are the longest so they are the critical paths and to be calculated first.

The path C to finish has a free float equal to its total float by definition.

The path H to C has a free float of (14-12-1=) 1. Path H to I has a free float of (14-12-1=) 1. Path H to F has a free float of (13-12-1=) 0. (Task H is generally held to have a free float of the minimum of these three).

The path F to finish has a free float equal to its total float by definition and doesn't have to be calculated again. The path E to I has a free float of (14-12-1=) 1. Path E to F has a free float of (13-12-1=) 0. (Task E is generally held to have a free float of the minimum of these two).

Question 40 - Develop the CPM chart (Very Hard)

Question:

Your project consists of nine tasks, A, B, C, D, E, F, G, H, and I. Task A is expected to take 10 days, task B is expected to take 21 days, task C is expected to take 16 days, task D is expected to take 12 days, task E is expected to take 19 days, task F is expected to take 21 days, task G is expected to take 18 days, task H is expected to take 16 days, and task I is expected to take 20 days. Task A, D, and G begin concurrently. Task B is preceded by task A and G. Task C is preceded by task B and task H and also ends the path. Task D precedes task E and task H. Task F is preceded by task E and H and is the last task on the path. Task H is preceded by task G, task A, and task D. Task I is preceded by task H, task B, and task E. It also ends the path. Diagram the project using PDM, then prepare the CPM chart, identify the critical path, calculate all the dates, and calculate the floats.

Answer:

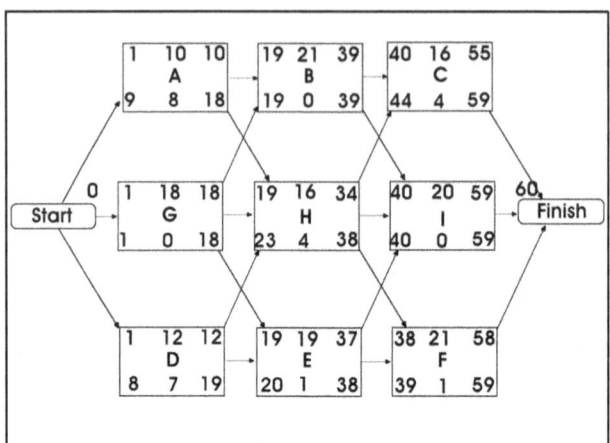

Process:

1. Design chart, placing letter or process name in the middle, and the duration on the top row in the middle.

2. There are seventeen paths A-B-C, D-E-F, G-H-I, A-B-I, A-H-C, A-H-I, A-H-F, G-B-C, G-B-I, G-H-C, G-H-F, G-E-I, G-E-F, D-H-C, D-H-I, D-H-F, and D-E-I. A-B-C has a total duration of 10+21+16=47. G-H-I has a total duration of 18+16+20=54. D-E-F has a total duration of 12+19+21=52. A-B-I has a total duration of 10+21+20=51. A-H-C has a total duration of 10+16+16=42. A-H-I has a total duration of 10+16+20=46. A-H-F has a total

192

duration of 10+16+21=47. G-B-C has a total duration of 18+21+16=55. G-B-I has a total duration of 18+21+20=59. G-H-C has a total duration of 18+16+16=50. G-H-F has a total duration of 18+16+21=55. G-E-I has a total duration of 18+19+20=57. G-E-F has a total duration of 18+19+21=58. D-H-C has a total duration of 12+16+16=44. D-H-I has a total duration of 12+16+20=48. D-H-F has a total duration of 12+16+21=49. D-E-I has a total duration of 12+19+20=51. G-B-I is the longest so it is the critical path and to be calculated first.

3. Calculate the forward pass for the critical path G-B-I. This goes on the upper row of tasks G, B, and I.

4. Perform the backward pass on the critical path. To do this, copy the Early Start and Early Finish from the upper row of tasks G, B, and I into the Late Start and Late Finish. The Late Start and Late Finish go into the bottom row of tasks G, B, and I.

5. Calculate the floats for the critical path. This is always zero and is placed in the middle

position of the bottom row of tasks G, B, and I. The free float is also zero and therefore is ignored.

6. Calculate the forward paths for the non-critical tasks A and C. This is placed on the top row of A and C.

Calculate the forward paths for the non-critical task H. This is placed on the top row of H.

Calculate the forward paths for the non-critical tasks D, E and F. This is placed on the top row of D, E, and F.

7. Calculate the backward paths for the non-critical tasks A and C. This is placed on the bottom row of tasks A, and C.

Calculate the backward paths for the non-critical tasks H. This is placed on the bottom row of tasks H.

Calculate the backward paths for the non-critical tasks D, E, and F. This is placed on the bottom row of tasks D, E, and F.

8. Calculate the total float for the non-critical tasks A, C, D, E, F, and H. This is placed in the middle position of the bottom row of tasks A, C, D, E, F, and H.

9. The path C to finish has a free float equal to its total float by definition and doesn't have to be calculated again. A to B has a free float of (19-10-1=) 8. Path A to H has a free float of (19-10-1=) 8. (Task A is generally held to have a free float of the minimum of these two).

The path H to C has a free float of (40-34-1=) 5. Path H to I has a free float of (40-34-1=) 5. Path H to F has a free float of (38-34-1=) 3. (Task H is generally held to have a free float of the minimum of these three).

The path F to finish has a free float equal to its total float by definition and doesn't have to be calculated again. The path E to I has a free float of (40-37-1=) 2. Path E to F has a free float of (38-37-1=) 0. (Task E is generally held to have a free float of the minimum of these two). Path D to H has a free float of (19-12-1=) 6. Path D to E has a free float (19-12-1=) 6. (Task D is generally held to have a free float of the minimum of these two).

Question 41 - Interpret the CPM chart (Very Hard)

Scenario:

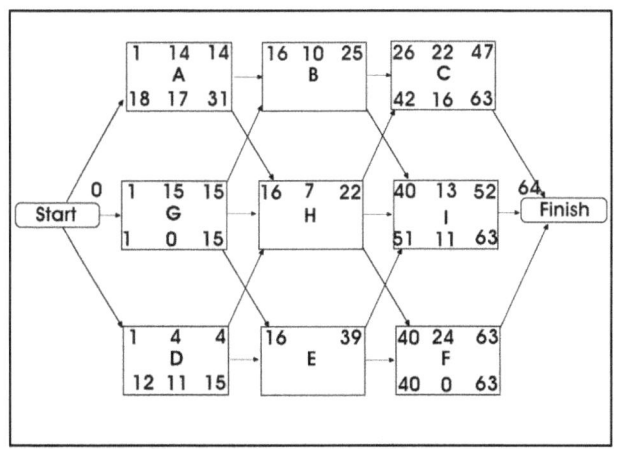

Question:

Given the above information, calculate the following using the minimum effort:

1. What is the total float for task E?
2. What is the duration for task E?
3. What is the late finish for task H?
4. What is the total float for task H?

Check your results by calculating all missing information.

Answer:

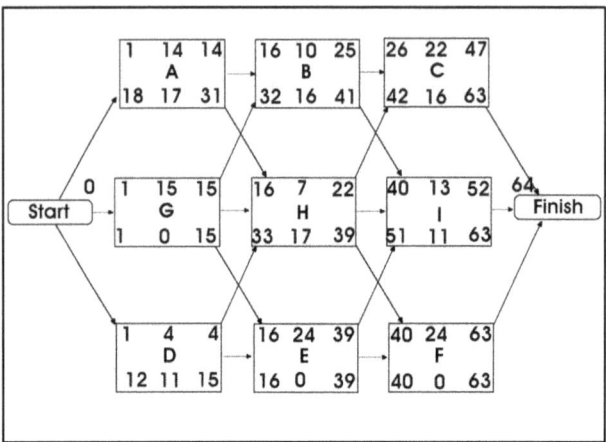

The answers are as shown in the above diagram.

To obtain the answers to the questions:

1. One of B, H, and E must have a total float of 0. E is the only one that extends for the whole range 16 to 39. Therefore E must be the critical path item and therefore total float must be 0.
2. Early Finish minus Early Start plus 1.

3. The lowest of Late Start for C, F, and I is 40. Therefore Late Finish for H must be one less.

4. Late Finish minus Early Finish.

Question 42 - Interpret the CPM chart (Very Hard)

Scenario:

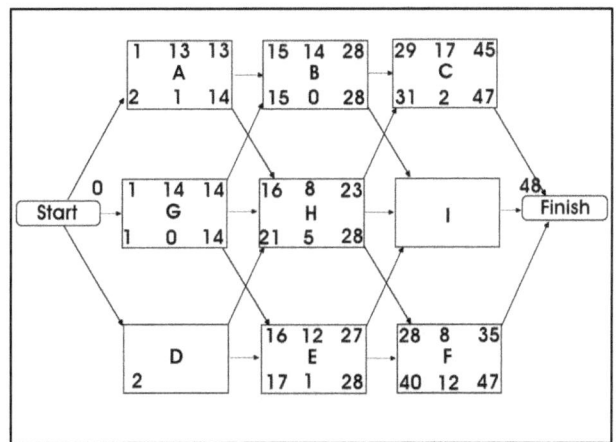

Question:

Given the above information, calculate the following using the minimum effort:

1. Early Finish for D
2. Duration for D
3. Early Start for I
4. Free Float for I

Check your results by calculating all missing information.

Answer:

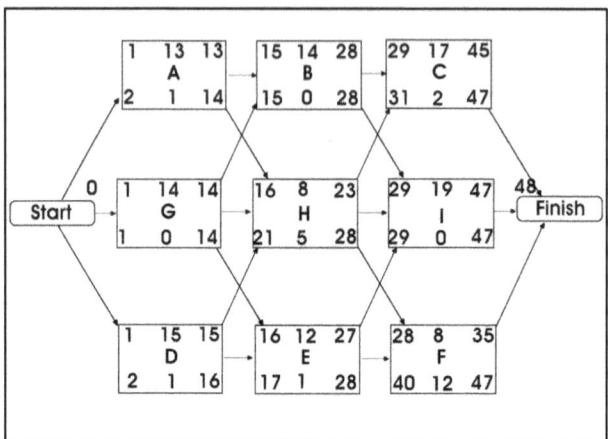

The answers are as shown in the above diagram.

To obtain the answers to the questions:

1. Early Start for H and E is greater than Early Finish for G plus one. Therefore Early Finish for D is one less than the Early Start for H and E.

2. Duration equals Early Finish for the first task in the path.

3. Greater of Late Finish for B, H, or E plus 1.

4. Neither C nor F are on the critical path. Therefore I must be. If I is on the critical path, then the Free Float (and Total Float) must be 0.

Question 43 - Interpret the CPM chart (Very Hard)

Scenario:

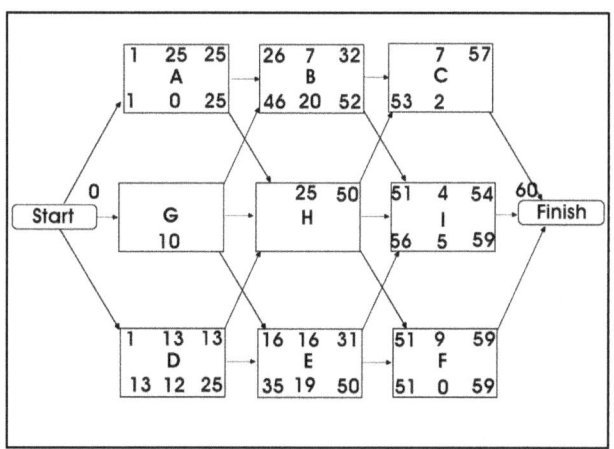

Question:

Given the above information, calculate the following using the minimum effort:

1. Late Finish of H
2. Late Start of H
3. Late Start of G
4. Duration of G

Check your results by calculating all missing information.

Answer:

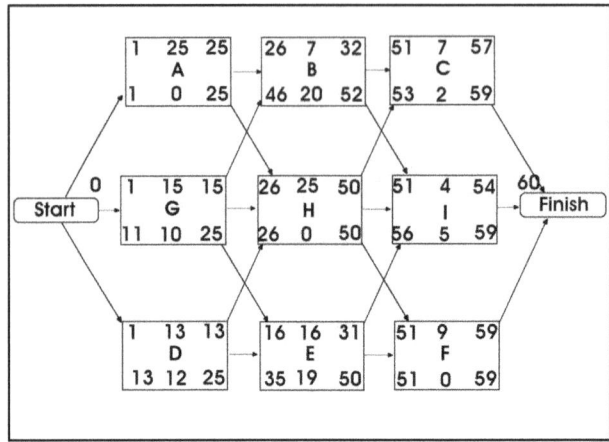

The answers are as shown in the above diagram.

To obtain the answers to the questions:

1. Neither B nor E are on the critical path. Therefore H must be. If H is on the critical path then Late Finish of H equals the Early Finish.

2. A and H are on the critical path. Therefore the Early Start of H is Early Finish of A plus one. The Late Start of H is the same as the

Early Start because H is on the Critical Path.

3. Early Start of G must be 1. The Total Float of G is 10. Therefore the Late Start of G is Early Start plus Total Float or 11.

4. Late Finish of G is the least of the Late Starts of B, H, and I less one. Duration is Late Finish less Late Start plus 1.

Question 44 - Interpret the CPM chart (Very Hard)

Scenario:

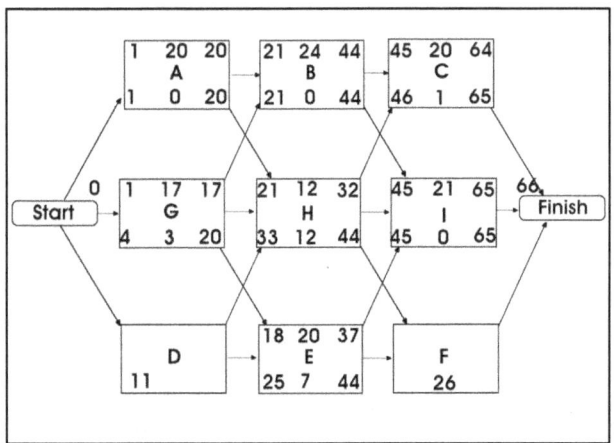

Question:

Given the above information, calculate the following using the minimum effort:

1. Late Start of F
2. Duration of F
3. Total Float of D
4. Early Finish of D

Check your results by calculating all missing information.

Answer:

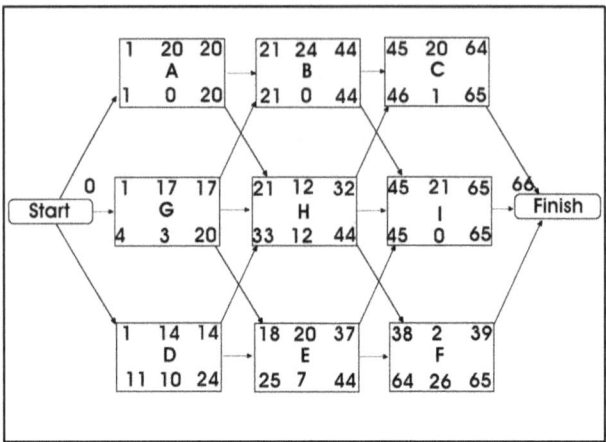

The answers are as shown in the above diagram.

To obtain the answers to the questions:

1. Early Start of F is the later of the Late Finish for E and H plus one. Late Start is Early Start plus Total Float.

2. Late Finish dates for all tasks succeeded by the Finish are the same. Duration of F is the Late Finish less the Late Start.

3. Early Start for D is 1 because it is the first task. Total Float is the Late Start less the Early Start.
4. Late Finish of D is the lesser of the Late Start for H and E less one. Early Finish is the Late Finish less the Total Float.

Question 45 - Interpret the CPM chart (Very Hard)

Scenario:

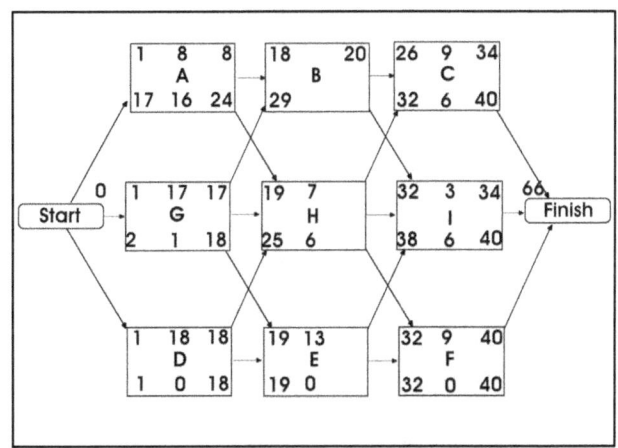

Question:

Given the above information, calculate the following using the minimum effort:

1. Late Finish for H
2. Early Finish for H
3. Late Finish for B
4. Late Finish for E

Check your results by calculating all missing information.

Answer:

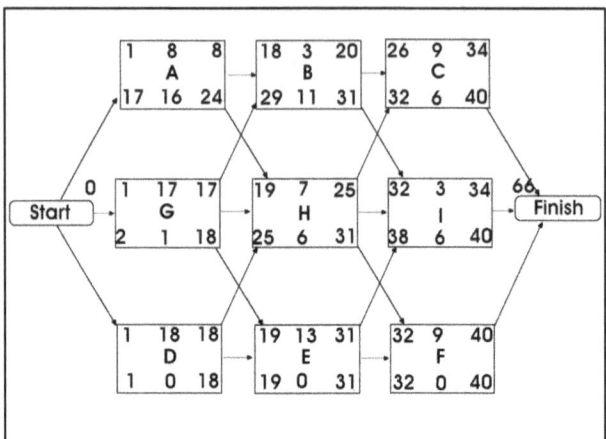

The answers are as shown in the above diagram.

To obtain the answers to the questions:

1. The Late Finish for H is the lesser of the Late Starts for C, I and F less one.
2. The Early Start for C is one more than the greater of the Early Finish for B and H. Since the Early Finish of B is six less than the Early Start of C, The Early Start of C must be based on the Early Finish of H.

Therefore the Early Finish of H is one less than the Early Start of C.

3. Late Finish for B is one less than the lessor of the Late Starts of C and I.

4. Late Finish for E is one less than the lessor of the Late Starts of F and I.

Question 46 - Interpret the CPM chart (Very Hard)

Scenario:

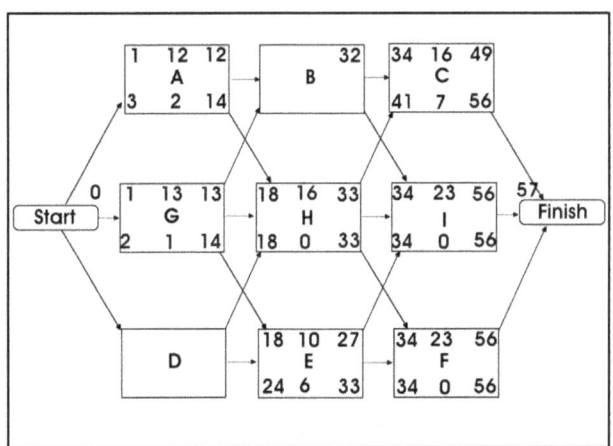

Question:

Given the above information, calculate the following using the minimum effort:

1. Total Float for D
2. Duration for D
3. Early Start for B
4. Total Float for B

Check your results by calculating all missing information.

Answer:

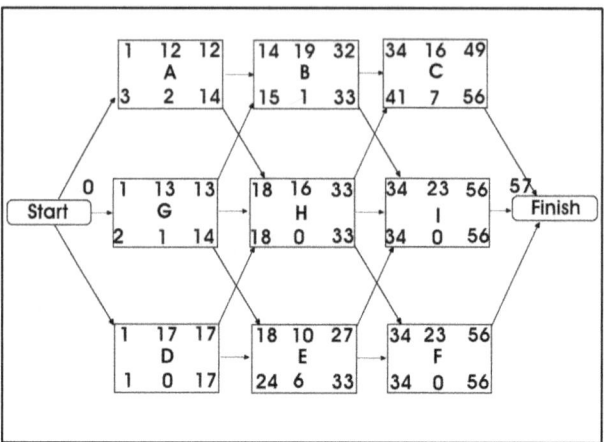

The answers are as shown in the above diagram.

To obtain the answers to the questions:

1. One of A, G, or D must be in the Critical Path. A & G are not, therefore D must be. If D is in the Critical Path then the Total Float for D is 0.

2. D is the first activity in the Critical Path. The Duration and Finish Dates for D equal 1 less than the minimum Early Start Date for H and E.

3. Early Start for B is one more than the maximum Early Finish Date for A and G.

4. The Late Start date for B is one more than the maximum Late Finish date for A and G. The Total Float for B is the difference between the Late Start date and the Early Start date for B. (See 3 to calculate the Early Start.)

Question 47 - Interpret the CPM chart (Very Hard)

Scenario:

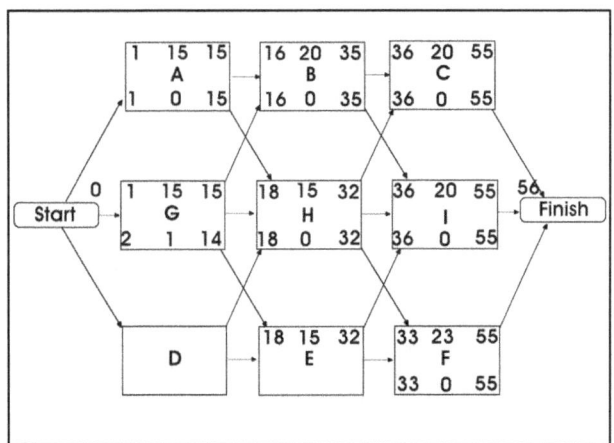

Question:

Given the above information, calculate the following using the minimum effort:

1. Late Finish for E
2. Total Float for E
3. Total Float for D
4. Duration for D

Check your results by calculating all missing information.

Answer:

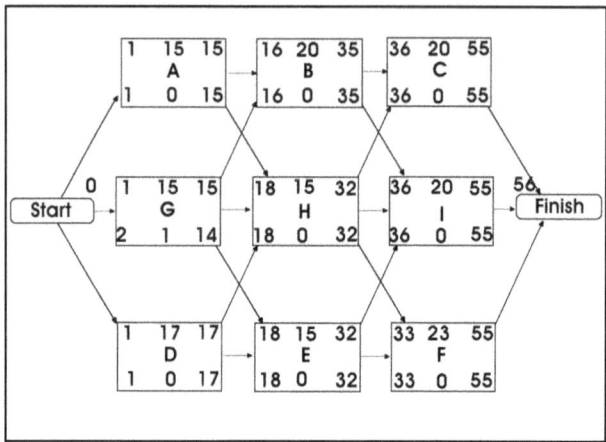

The answers are as shown in the above diagram.

To obtain the answers to the questions:

1. Late Finish for E equals one less than the minimum Late Start for I and F.
2. Total Float for E equals Late Finish for E less Late Start for E.
3. For E to be on the Critical Path (Total Float = 0) then either G or D is a critical path activity. G is not therefore

D is. Therefore the Total Float for D is 0.

4. Early Finish for D is the minimum of the Early Start of H and E less one. Because D is the first task on a path, Duration for D is the Early Finish for D.

Question 48 - Interpret the CPM chart (Very Hard)

Scenario:

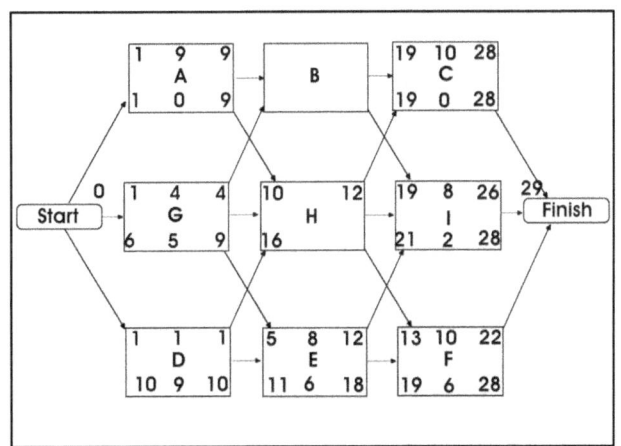

Question:

Given the above information, calculate the following using the minimum effort:

1. Early Start for B
2. Total Float for B
3. Late Finish for H
4. Total Float for H

Check your results by calculating all missing information.

Answer:

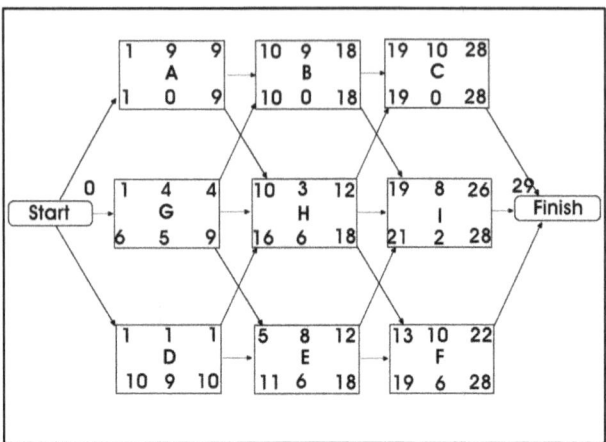

The answers are as shown in the above diagram.

To obtain the answers to the questions:

1. Early Start for B is one more than the maximum for the Early Finish of A and G.

2. A and C are on the Critical Path. Therefore either B or H must be on the Critical Path. H is not, therefore B must be. Total Float for B is 0.

3. Late Finish for H is one less than the minimum Late Start for C, I, and F.

4. Total Float for H is Late Start for H minus Early Start for H. (Sometimes the basic calculation is the easiest method).

Question 49 - Interpret the CPM chart (Very Hard)

Scenario:

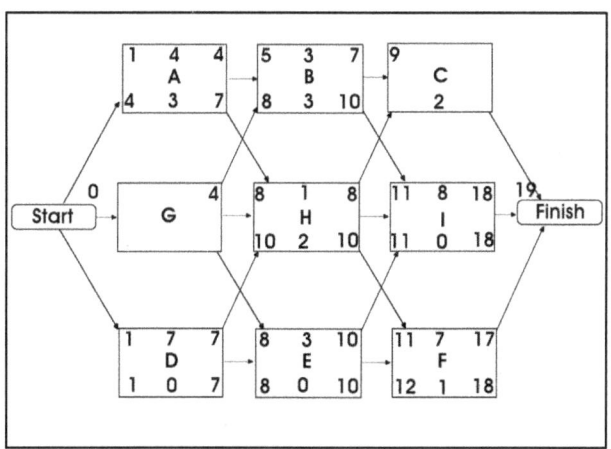

Question:

Given the above information, calculate the following using the minimum effort:

1. Early Finish for C
2. Late Start for C
3. Duration for G
4. Late Start for G

Check your results by calculating all missing information.

Answer:

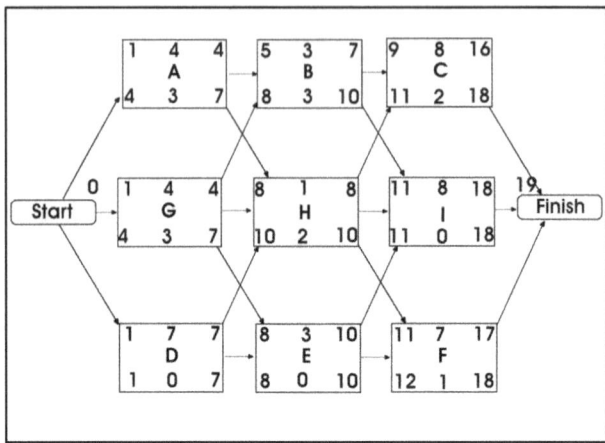

The answers are as shown in the above diagram.

To obtain the answers to the questions:

1. Late Finish for C is one less than the overall Finish date. Or the same as all the other Late Finish dates for C, I, and F (i.e. last in the path). Early Finish date for C is Late Finish less the Total Float.
2. Late Start for C is the sum of Early Start and Total Float for C.

3. Duration for G is the same as the Early Finish for G.

4. Late Finish for G is one less than the minimum Late Start for B, H, and E. Late Start for G is the Late Finish less the Duration plus one.

Question 50 - Interpret the CPM chart (Very Hard)

Scenario:

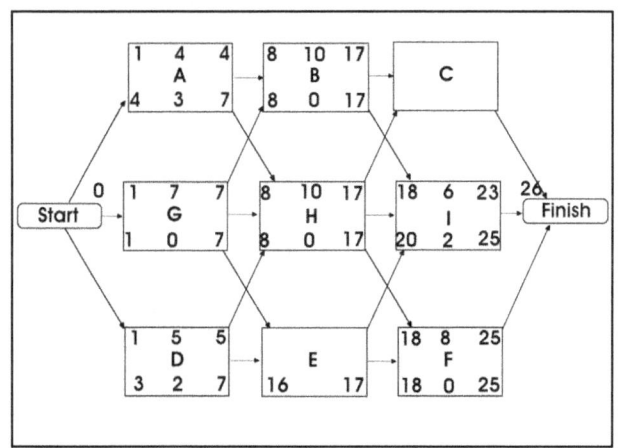

Question:

Given the above information, calculate the following using the minimum effort:

1. Total Float of C
2. Early Finish of C
3. Early Start of E
4. Duration of E

Check your results by calculating all missing information.

Answer:

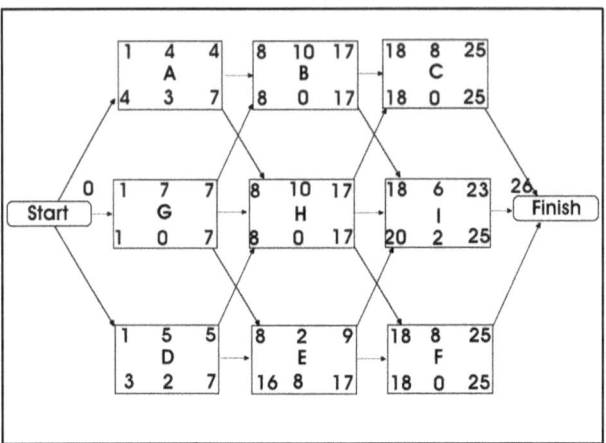

The answers are as shown in the above diagram.

To obtain the answers to the questions:

1. Late Finish of B, H and I are identical. The Late Start of I is more than one greater than the Late Finish of B, H and I. Therefore the Late Start of C must be one greater than the Late Finish of B, and H. The Late Finish of C is the last period before the Finish date. Therefore C takes up the entire

period available from B to Finish. That puts it on the Critical Path. Total Float of C is therefore 0.

2. Because the Total Float is zero the Early Finish of C is one less than the Finish date or the Late Finish.

3. Early Start of E is one greater than the maximum Early Finish of G and D.

4. Duration of E is the difference between Late Finish and Late Start plus one.

Chapter 4: A final note and an offer

I hope you enjoyed this book and that you find it useful in your pursuit of the PMP® designation. As I mentioned before, I would very much appreciate your reviewing this book on the website where you purchased it. If there is no review facility there, or if you purchased it from a bricks and mortar store, please review it on Amazon and/or Goodreads.

Reviews are very important to independent authors and small press authors like myself. The effect on our sales can be enormous. Not only do they improve sales and placement for our web promotion, but the feedback is critical for us to improve. Be nice to an author and review their book! We get lonely.

In addition, at the time I wrote and published this book, I didn't have a giveaway. However, by the time you buy this book, there will be a free gift for my readers at http://GlenDFord.com. Come by and claim your bonus gift. The coupon code will be

CPMACE

I don't know what the gift will be but it will be worth it. You can also join my email list there and keep in touch.

Enjoy and Good Luck.

Glen Ford
Mississauga, Ontario, Canada
February, 2017

Glen Ford

Glen Ford was formerly the Chief Operating Officer and Chief Information Officer as well as a co-founder with VProz Inc. He is a serial entrepreneur having set up the internet training company TrainingNOW and its subsidiaries as well as providing consulting services for startups in Debt Counseling, Software and Payment Processing. He has been principal of his own project management consultancy for over 11 years. During that time he has alternated his clients between government, the big banks and small to medium companies. Prior to that he spent 10 years working for the Canadian Standards Association and 10 years alternating between large distribution and manufacturing companies. He also worked for a very successful HVAC firm. Glen is now training, writing, coaching, and consulting on project management and related entrepreneurship topics including the implementation of PMOs and methodologies. You can reach him directly through his website http://www.GlenDFord.com.

Glen is active in the business community as a member of The Project Management Institute (PMI) Lakeshore Chapter and a former training director for BNI Eagles Chapter of Business Network International (BNI). Glen is also an active supporter of charity including Scouts Canada (3rd Erin Mills Scouts). Glen holds a BSc from McMaster University, an MCPM from York University (Schulich), and a PMP (Project Management Professional) designation.

TrainingNOW *Training*NOW

TrainingNOW is a training and publishing company located in Mississauga Ontario, Canada and your computer screen. It provides specialized web hosting services for companies seeking to deliver "how to" education over the web. It also publishes and sells "how to" books and training materials in digital, print and other media. Through its subsidiaries LearningCreators and ContentCreators it provides training on how to write your own book and, provides custom training material development including books.

Glen Ford is available as a trainer, speaker, coach, or consultant. You can find more information on his courses, and services at http://GlenDFord.com

You can find more information on training courses, books, and publishing your own books at http://TrainingNOW.ca

You can email the writer through either site.

If you liked this book ...

For More Books On This and Similar Topics
http://www.trainingnow.ca

Books By Glen Ford

On Project Management and Business

How to Document a Project Plan: What you need to know to design a project management plan quickly and easily

If you are setting up a PMO, consulting, leading or project managing in an organization without a methodology for project management, you need this book. It is a comprehensive guide to all the forms you will need to document your projects. Never lose track of your workload again.

Ace the PMP® Exam: 50 Critical Path Method
(CPM) exercises to help you pass your PMP exam

Are you writing your Project Management Professional exam? If so you need this book. Being able to calculate CPM by hand is a critical part of the knowledge you need to pass. But almost no one actually calculates CPM charts by hand anymore. How can you be sure you know it well enough to pass? This book will help give you that confidence.

Ace the PMP® Exam: 50 Earned Value Management (EVM) exercises to help you pass your PMP exam

Are you writing your Project Management Professional exam? If so you need this book. Being able to calculate the EVM formulas is a critical part of the knowledge you need to pass. But almost no one does EVM. How can you be sure you know it well enough to pass? This book will help give you that confidence.

How to Blog for Money: 9 strategies to get your blog earning money online and off

You've heard all the hype about earning big money with your blog. Everyone seems to want to sell you a course on the topic but no one seems to be very open about actually helping you understanding how to do it. Well, this book breaks that pattern. In this book, you'll learn the nine models for actually making money with your blog, the advantages of each, and the downsides of each.

For More Information On This And Related Topics
http://www.GlenDFord.com/blog

On writing

How to Write Your Own How-To EBook in 24 Hours or Less: The information products secret revealed!

Would you like to write your own non-fiction book? Non-fiction books are a great way to market your business or prove your expertise. A book is still the best way to establish yourself as an expert in your field. But it can be daunting. It takes so long to write. And it's so expensive. And what if people don't like it? It might end up hurting your reputation if it looks cookie-cutter. This book solves those problems. You can create a very high quality non-fiction book quickly and easily. One that will help your reputation (not hurt it). A book you can be proud of.

Writer's Block Demolition: Finding the time to write, keeping writing, and finish YOUR book

Are you having a problem finishing your book? Not enough time? Have you tried the other books on writer's block and they just haven't worked? Even worse, have their cookie-cutter solutions left you feeling gun-shy and despairing of ever writing? There's a reason why those other books failed. Until you determine exactly why you can't write, you'll never eliminate the problem. This book will guide you through determining why you are having problems, explain the three things you need to write, and give you practical solutions to help you make sure you have those three things in your life. And much, much more.

101 Writing Tweets: 101 tips and tweets about writing how-to books for the Kindle

Are you looking to make money writing books? Would you like some tips but don't want to read a big book to pull out only a few bits of gold? Here are 101 tweets on writing. Quick tips in 140 characters. Quick to read, easy to understand, and each one a gem. Plus, there's an explanation of the tip beneath each tweet. Read one a day and learn more about the business of writing.

For More Information On This And Related Topics
http://www.learningcreators.com/blog

As Glen Douglas

How To Build A Raised Garden Bed

Raised garden beds are a practical solution to many problems. Have problems with bending? Have problems with animals attacking your vegetables? Want to separate your plantings? Want different soil conditions for your plants? Raised garden beds solve all these problems and look beautiful doing it. And they are easy to build with this book.

With Paul Benson

101 Limericks About Public Speaking

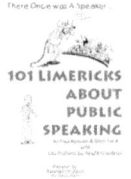

Did you know that many people are more afraid of speaking in public than they are of death? It's true. But one of the best ways to overcome fear is to laugh at it. In this book, we matched 101 limericks about the silly side of public speaking with 101 funny pictures by renowned animator Aputik. We laugh. We teach. And fear fades away.

7 Myths and Seven Tricks in Nine Steps: The truth & tricks about learning course product creation that THEY don't know

Have you heard the gurus telling you to start a business selling your knowledge? Courses are everywhere. It's easy! It's profitable! It's ... well, there's myths and half-truths everywhere you look. This book cuts through the nonsense.

Most books are available in both Kindle and Print versions.

Audio and Video Courses

How to Write Your Own Book or eBook in 24 hours or Less! **FREE** on-line course with Glen Ford. Do you want to write a how-to book? There are many reasons you might want to. If you are self-employed, books are a great way to distinguish yourself and build a brand as an expert in your field. If you are a knowledge worker, a book is one of the five cornerstones in your personal branding efforts. If you just have something to say, a desire to get your message out, writing a book is the best way to get your message out there. In this **FREE course**, you'll learn a method that has been tested, tried, and used by hundreds of education professionals for over thirty years. Just updated and enhanced to be easier to use and better suited to your needs. This course introduces the same method described in Glen Ford's book How to Write Your Own Book or eBook in 24 hours or Less!

 Researching Your Book – How To Make Your Next Book Sizzle! audio program with Paul Benson. Can you really afford to write a book no one wants? Or you've written one course and now where do you go? Sometimes you just have to do research. This audio program shows you how to do just enough research and no more.

 Finding the Time To Write: Time Management for Writers video program with Glen Ford. Are you still sitting there hoping to write your book someday? Do you keep telling yourself "Tomorrow I'll start." but

tomorrow never comes? Are you afraid to start? The truth is there are three different problem areas that you need to resolve if you are going to get writing. In this 2 DVD set you'll be walked through a method that will get you to sit down and write. This was originally a workshop and it is meant to walk you through the process from fear to fantastic! Writer's Block Demolition: Finding the Time to Write, Keeping Writing, and Finish Your Book is a great resource for this course.

www.ingramcontent.com/pod-product-compliance
Lightning Source LLC
Chambersburg PA
CBHW051635170526
45167CB00001B/196